AVIATION SECURITY

FOREIGN AIRPORTS
AND INBOUND AIR CARGO

TRANSPORTATION ISSUES, POLICIES AND R&D

Additional books in this series can be found on Nova's website under the Series tab.

Additional E-books in this series can be found on Nova's website under the E-book tab.

DEFENSE, SECURITY AND STRATEGIES

Additional books in this series can be found on Nova's website under the Series tab.

Additional E-books in this series can be found on Nova's website under the E-book tab.

TRANSPORTATION ISSUES, POLICIES AND R&D

AVIATION SECURITY

FOREIGN AIRPORTS AND INBOUND AIR CARGO

JUN BAI
EDITOR

New York

For permission to use material from this book please contact us:
Telephone 631-231-7269; Fax 631-231-8175
Web Site: http://www.novapublishers.com

NOTICE TO THE READER

The Publisher has taken reasonable care in the preparation of this book, but makes no expressed or implied warranty of any kind and assumes no responsibility for any errors or omissions. No liability is assumed for incidental or consequential damages in connection with or arising out of information contained in this book. The Publisher shall not be liable for any special, consequential, or exemplary damages resulting, in whole or in part, from the readers' use of, or reliance upon, this material. Any parts of this book based on government reports are so indicated and copyright is claimed for those parts to the extent applicable to compilations of such works.

Independent verification should be sought for any data, advice or recommendations contained in this book. In addition, no responsibility is assumed by the publisher for any injury and/or damage to persons or property arising from any methods, products, instructions, ideas or otherwise contained in this publication.

This publication is designed to provide accurate and authoritative information with regard to the subject matter covered herein. It is sold with the clear understanding that the Publisher is not engaged in rendering legal or any other professional services. If legal or any other expert assistance is required, the services of a competent person should be sought. FROM A DECLARATION OF PARTICIPANTS JOINTLY ADOPTED BY A COMMITTEE OF THE AMERICAN BAR ASSOCIATION AND A COMMITTEE OF PUBLISHERS.

Additional color graphics may be available in the e-book version of this book.

Library of Congress Cataloging-in-Publication Data

ISBN: 978-1-62618-048-2

Published by Nova Science Publishers, Inc. † *New York*

CONTENTS

PREFACE

International flights bound for the United States continue to be targets of terrorist activity, as demonstrated by the October 2010 discovery of explosive devices in air cargo packages bound for the United States from Yemen. The Transportation Security Administration (TSA) is responsible for securing the nation's civil aviation system, which includes ensuring the security of U.S.-bound flights. This book explores the steps the TSA has taken to enhance its foreign airport assessment procedures since 2007, with a focus on foreign airports and inbound air cargo.

Chapter 1 – International flights bound for the United States continue to be targets of terrorist activity, as demonstrated by the October 2010 discovery of explosive devices in air cargo packages bound for the United States from Yemen. The Transportation Security Administration (TSA) is responsible for securing the nation's civil aviation system, which includes ensuring the security of U.S.-bound flights. As requested, GAO evaluated (1) the steps TSA has taken to enhance its foreign airport assessment program since 2007, and any remaining program challenges; (2) TSA's assessment results, including how TSA uses the results to guide future efforts; and (3) what opportunities, if any, exist to enhance the program. To conduct this work, GAO reviewed foreign airport assessment procedures and results, interviewed TSA and foreign aviation security officials, and observed TSA conduct a foreign airport assessment. While these interviews and observations are not generalizable, they provided insights on TSA's program. This is the public version of a sensitive report GAO issued in September, 2011. Information that TSA deemed sensitive has been omitted.

Chapter 2 - In 2010, passenger flights transported about 3.6 billion pounds of cargo into the United States from foreign locations. According to TSA, the introduction of explosive devices in air cargo shipments is a significant threat, and DHS was mandated to establish a system to screen 100 percent of cargo transported on all passenger aircraft traveling to, from, or within the United States by August 2010. Individuals identified as matches to the No Fly List are generally prohibited from boarding commercial aircraft because it has been determined they pose a threat to civil aviation or national security. GAO was asked to examine (1) TSA actions taken since October 2010 to enhance the security of inbound air cargo transported on both passenger aircraft and all-cargo carriers; and (2) any associated challenges TSA faces. GAO reviewed relevant requirements and documents, interviewed federal officials, and visited three airports based on cargo volume. The visits provided insights, but were not generalizable to the entire industry. This is a public version of a sensitive security report GAO issued in March 2012, which also addressed U.S. Customs and Border Protection's and TSA's use of the No Fly List to secure inbound air cargo.

In: Aviation Security
Editor: Jun Bai

ISBN: 978-1-62618-048-2
© 2013 Nova Science Publishers, Inc.

Chapter 1

AVIATION SECURITY: TSA HAS TAKEN STEPS TO ENHANCE ITS FOREIGN AIRPORT ASSESSMENTS, BUT OPPORTUNITIES EXIST TO STRENGTHEN THE PROGRAM[*]

United States Government Accountability Office

WHY GAO DID THIS STUDY

International flights bound for the United States continue to be targets of terrorist activity, as demonstrated by the October 2010 discovery of explosive devices in air cargo packages bound for the United States from Yemen. The Transportation Security Administration (TSA) is responsible for securing the nation's civil aviation system, which includes ensuring the security of U.S.-bound flights. As requested, GAO evaluated (1) the steps TSA has taken to enhance its foreign airport assessment program since 2007, and any remaining program challenges; (2) TSA's assessment results, including how TSA uses the results to guide future efforts; and (3) what opportunities, if any, exist to enhance the program. To conduct this work, GAO reviewed foreign airport assessment procedures and results, interviewed TSA and foreign aviation security officials, and observed TSA conduct a foreign airport assessment.

---

[*] This is an edited, reformatted and augmented version of the Highlights of GAO-12-163 a report to congressional requesters, dated October 2011.

While these interviews and observations are not generalizable, they provided insights on TSA's program. This is the public version of a sensitive report GAO issued in September, 2011. Information that TSA deemed sensitive has been omitted.

WHAT GAO RECOMMENDS

GAO recommends that TSA develop a mechanism to evaluate its assessment results to identify any trends, and target resources and future activities; establish criteria for determining foreign airport vulnerability ratings; and consider the feasibility of conducting more targeted assessments and compiling information on aviation security best practices. DHS agreed with the recommendations.

WHAT GAO FOUND

Since 2007, TSA has taken a number of steps to enhance its foreign airport assessment program, some of which were taken in response to GAO's prior recommendations. For example, TSA updated its policies and methodologies used to guide and prioritize its assessment efforts, and implemented tools to track its annual assessment schedule, airport assessment results, and foreign government progress in resolving security deficiencies previously identified during the assessments. However, challenges remain in gaining access to some foreign airports, developing an automated database to better manage program information, prioritizing and providing training and technical assistance to foreign countries, and expanding the scope of TSA's airport assessments to include all- cargo operations. TSA has various efforts under way to address these challenges.

Based on GAO's analysis of TSA's foreign airport assessments conducted from fiscal year 2006 through May 2011, some foreign airports complied with all of TSA's aviation security assessment standards; however, TSA has identified serious noncompliance issues at a number of foreign airports. Common areas of noncompliance included weaknesses in airport access controls and passenger and baggage screening. Moreover, GAO's analysis showed variation in airport compliance across geographic regions and individual security standards, among other things. For example, GAO's analysis showed that some number of regions of the world had no airports with egregious noncompliance while other regions had several such airports. However, TSA has not yet taken steps to evaluate its assessment results to identify regional and other trends over time. Developing a mechanism to evaluate its assessment results could help support TSA's priorities for aviation security training and technical assistance, inform its risk management decision making by identifying any trends and security gaps, and target capacity building efforts.

Opportunities also exist for TSA to make additional program improvements in several key areas. For example, the agency has not developed criteria and guidance for determining foreign airport vulnerability ratings. This is particularly important given that these ratings are a key component for how TSA determines each foreign airport's risk level. Providing TSA decision makers with more specific criteria and definitions could provide greater assurance that such determinations are consistent across airports over time. In addition, there are opportunities for TSA to increase program efficiency and effectiveness by, for example, conducting more targeted foreign airport assessments and systematically compiling and analyzing security best practices. Taking such actions could help TSA better focus its assessments to address areas of highest risk, and identify security best practices and technologies that may be applicable to enhancing the security of both foreign and domestic airports.

ABBREVIATIONS

AIT	Advanced Imaging Technology
AME	Africa-Middle East
APAC	Asia-Pacific
ASSIST	Aviation Security Sustainable International Standards Team
ATA	Anti-Terrorism Assistance
CDB	Capacity Development Branch
DHS	Department of Homeland Security
EC	European Commission
ECAC	European Civil Aviation Conference
EU	European Union
EUR	Europe
FAA	Federal Aviation Administration
FAARS	Foreign Airport Assessment Reporting System

GC	Global Compliance
GPRA	Government Performance and Results Act
ICAO	International Civil Aviation Organization
OGS	Office of Global Strategies
OIO	Office of International Operations
ROC	Regional Operations Center
SOP	Standard Operating Procedures
SSI	Sensitive Security Information
TSA	Transportation Security Administration
TSAR	TSA Representative
WH	Western Hemisphere

October 21, 2011

The Honorable Susan M. Collins
Ranking Member
Committee on Homeland Security and Governmental Affairs
United States Senate

The Honorable Peter T. King
Chairman
Committee on Homeland Security
House of Representatives

The Honorable John L. Mica
Chairman
Committee on Transportation and Infrastructure
House of Representatives

The December 25, 2009, attempt to detonate an explosive during an international flight bound for Detroit, and the October 2010 discovery of explosive devices in air cargo packages bound for the United States from Yemen provide vivid reminders that civil aviation remains a key terrorist target and highlight the importance of ensuring the security of U.S.-bound flights. Furthermore, roughly 80 million passengers and 10 billion pounds of cargo are transported on inbound flights to the United States per year, further highlighting the need to ensure the security of these flights. Approximately 300 foreign airports provide last point of departure flights to the United States in approximately 100 countries. As a result, efforts to evaluate the security of foreign airports that service the United States— and mitigating any identified security risks—are important steps in ensuring the security of the U.S. aviation system particularly considering that inbound flights continue to be targets of coordinated terrorist activity.

The Transportation Security Administration (TSA), a component of the Department of Homeland Security (DHS), is the federal agency with primary responsibility for securing the nation's civil aviation system, which includes ensuring the security of U.S.-bound flights.[1] Through its foreign airport assessment program, TSA determines whether foreign airports that provide service to the United States are maintaining and carrying out effective security measures.[2] While TSA is authorized under U.S. law to conduct foreign airport assessments at intervals it considers necessary, TSA may not perform an assessment of security measures at a foreign airport without permission from the host government.

In 2007, we reported on TSA's efforts to assess the security at foreign airports and recommended that TSA, among other things, take steps to track the status of foreign airport assessments from initiation through completion, develop a standard process for tracking and documenting host governments' progress in addressing security deficiencies identified during TSA assessments, and develop outcome-oriented performance measures to evaluate the impact TSA assessments have on improving foreign airport compliance with international standards.[3] DHS concurred with the recommendations and has since taken a number of steps to address them and improve the program. We discuss these and other actions TSA has taken, since 2007, later in this report.

Considering the high volume of passengers and flights arriving in the United States from foreign locations and the recent history of terrorist threats against inbound commercial flights, you asked us to reexamine TSA's foreign airport assessment program, including the progress it has made since our prior report in 2007. Specifically, this report addresses the following questions: (1) to what extent has TSA taken steps to enhance its foreign airport assessment program since 2007, and what challenges remain; (2) what are the results of TSA's foreign airport assessments, and to what extent does TSA use the results of these assessments to guide its future assessment activities; and (3) what opportunities, if any, exist to enhance the value of TSA's foreign airport assessment program?

This report is a public version of the prior sensitive report that we provided to you in September 2011. TSA deemed some of the information in the prior report as Sensitive Security Information (SSI), which must be protected from public disclosure. Therefore, this report omits sensitive information about the specific results of TSA's foreign airport assessments. In addition, at TSA's request, we have omitted some information regarding the remaining challenges for TSA's foreign airport assessment program. Although the information provided in this report is more limited in scope, it addresses the same questions as the sensitive report. Also, the overall methodology used for both reports is the same.

To collectively address these objectives, we obtained and reviewed TSA guidance for conducting and reporting the results of foreign airport assessments, such as TSA's *Foreign Airport Assessment Program Standard Operating Procedures* (SOP) document, which prescribes program and operational guidance for assessing security measures at foreign airports. We also obtained and analyzed the results of TSA's foreign airport assessments from fiscal year 2006 through May 9, 2011, to determine the extent to which foreign airports complied with international aviation security standards, and assessed how TSA conducted follow-up on the results of these assessments. Specifically, we analyzed the frequency with which foreign airports complied with standards, such as passenger screening, baggage screening, and access controls, among others. We assessed the reliability of TSA's foreign airport assessment data and concluded that the data were sufficiently reliable for the purposes of our review. Among the steps we took to assess the reliability of TSA data were selecting a random sample of records from the program's vulnerability results tracking sheet and examining the corresponding assessment reports to identify any inconsistencies. We also reviewed the steps TSA takes to assign risk rankings to foreign airports as well as efforts to analyze its assessment results, and compared these efforts to *Standards for Internal Control in the Federal Government*.[4] In addition, we interviewed TSA program management officials located at TSA headquarters as well as visited four of the five TSA Regional Operations Centers (ROC) located in Miami, Los Angeles, Dallas, and Frankfurt, Germany, to interview TSA international inspector officials. We based our site visit selections on the number of available inspectors at each location and geographic dispersion. We also interviewed other federal and nonfederal stakeholders, such as the Department of State, International Civil Aviation Organization (ICAO), and European Commission (EC) officials to discuss efforts and programs these organizations have in place to enhance international aviation security. In addition, we accompanied TSA officials during an assessment at a foreign airport to observe how TSA's policies and procedures were implemented in practice. Information from our interviews with government officials, members of the aviation industry, and TSA officials and inspectors, and our observations of TSA inspectors cannot be generalized beyond those that we spoke with because we did not use statistical sampling techniques in selecting individuals to interview. However, these interviews and observations provide perspectives on TSA's foreign airport assessment program, including various officials' roles and responsibilities related to the program.

We conducted this performance audit from August 2010 through October 2011 in accordance with generally accepted government auditing standards. Those standards require that we plan and perform the audit to obtain sufficient, appropriate evidence to provide a reasonable basis for our findings and conclusions based on our audit objectives. We believe that the evidence obtained provides a reasonable basis for our findings and conclusions based on our audit objectives. More details about the scope and methodology of our work are contained in appendix I.

BACKGROUND

DHS Responsibilities for Enhancing the Security of Airports with U.S.-Bound Flights from Foreign Countries

Shortly after the September 11, 2001, terrorist attacks, Congress passed, and the President signed into law, the Aviation and Transportation Security Act, which established TSA and gave the agency responsibility for securing all modes of transportation, including the nation's civil aviation system, which includes domestic and international commercial aviation operations.[5] In furtherance of its civil aviation security responsibilities, TSA is statutorily required to assess the effectiveness of security measures at foreign airports served by a U.S. air carrier, from which a foreign air carrier serves the United States, that pose a high risk of introducing danger to international air travel, and at other foreign airports deemed appropriate by the Secretary of Homeland Security.[6] This provision of law also identifies measures that the Secretary must take in the event that he or she determines that an airport is not maintaining and carrying out effective security measures based on TSA assessments.[7] See appendix II for a detailed description of the process for taking secretarial actions against a foreign airport. In addition, TSA conducts inspections of U.S. air carriers and foreign air carriers that service the United States from foreign airports pursuant to its authority to ensure that air carriers certified or permitted to operate to, from, or within the United States meet applicable security requirements, including those set forth in an air carrier's TSA-approved security program.[8]

The Secretary of Homeland Security delegated to the Assistant Secretary of TSA the responsibility for conducting foreign airport assessments, but retained responsibility for making the determination that a foreign airport does not maintain and carry out effective security measures. Currently, the Global Compliance Division and Office of International Operations, within TSA's Office of Global Strategies, are responsible for conducting foreign airport assessments.

Table 1 highlights the roles and responsibilities of the TSA positions within these divisions that are responsible for implementing the foreign airport assessment program.

Table 1. Positions That Play a Key Role in TSA's Foreign Airport Assessment Program

Office/division	Position	Duties
Global Compliance, Office of Global Strategies	Director of Global Compliance	The Director of Global Compliance carries out the statutory mandate of the Secretary of Homeland Security and the Assistant Secretary of TSA to assess the adequacy of civil aviation security at foreign airports. The Director of Global Compliance supervises and directs work of the Regional Operations Center Managers and assigned desk officers.
Global Compliance, Office of Global Strategies	Regional Operations Center (ROC) Manager	The ROC Manager is responsible for the overall planning and conduct of assessments of the foreign airports for which he/she has geographic responsibility, including the scheduling and coordination of personnel and resources. The ROC Manager supervises and directs the work of the inspector workforce and administrative support personnel within his/her assigned geographic responsibility.[a]
Global Compliance, Office of Global Strategies	Aviation Security Inspector	Inspectors are primarily responsible for performing and reporting the results of foreign airport assessments, and will provide on-site assistance and make recommendations for security enhancements. Inspectors are also deployed in response to specific incidents and to monitor for identified threats. Inspectors are based in one of TSA's five ROCs. As of July 2011, TSA has authorized 2,013 full-time equivalent inspectors, including 1,929 domestic aviation, cargo, cargo canine, and surface inspectors and 84 international aviation inspectors.

Office/division	Position	Duties
		Of the 84 authorized international aviation inspectors, 64 are on-board (with 10 vacancies) for international aviation inspectors and 9 are on-board (with 1 vacancy) for international cargo aviation inspectors.
International Operations, Office of Global Strategies	TSA Representative (TSAR)	TSARs communicate with foreign government officials to address transportation security matters and to conduct foreign airport assessments. Specifically, the TSARs serve as on-site coordinators for TSA responses to terrorist incidents and threats to U.S. assets at foreign transportation modes. TSARs also serve as principal advisors on transportation security affairs to U.S. ambassadors and other embassy officials responsible for transportation issues to ensure the safety and security of the transportation system. For the foreign airport assessment program, TSARs are often involved in arranging pre-assessment activities, assessment visits, and follow-up visits. Additionally, TSARs are responsible for completing portions of the airport assessment reports and reviewing completed assessment reports. TSARs also help host government officials address security deficiencies that are identified during assessments. As of July 2011, TSA had 24 TSARs.[b]

Source: TSA.

[a] TSA's ROCs are located in Dallas, Miami, Singapore, Los Angeles, and Frankfurt, Germany. They are responsible for foreign airports in the geographic regions of Africa-Middle East (AME), Asia- Pacific (APAC), Europe (EUR), and Western Hemisphere (WH).

[b] TSARs are located in Amman, Athens, Bangkok, Beijing, Berlin, Brussels, Buenos Aires, Johannesburg, Kabul, London, Madrid, Manila, Mexico City, Miami, Nairobi, Nassau, Ottawa, Paris, Rome, Singapore, Sydney, Tokyo, and Warsaw. There is also a TSA liaison to the United States Africa Command based in Stuttgart, Germany.

TSA's Process for Assessing Aviation Security Measures at Foreign Airports

TSA assesses the effectiveness of security measures at foreign airports using select aviation security standards and recommended practices adopted by ICAO, a United Nations organization representing 190 countries.[9] ICAO standards and recommended practices address operational issues at an airport, such as ensuring that passengers and baggage are properly screened and that unauthorized individuals do not have access to restricted areas of an airport. ICAO standards and recommended practices also address non-operational issues, such as whether a foreign government has implemented a national civil aviation security program for regulating security procedures at its airports and whether airport officials implementing security controls go through background investigations, are appropriately trained, and are certified according to a foreign government's national civil aviation security program. ICAO member states have agreed to comply with these standards, and are strongly encouraged to comply with ICAO-recommended practices.[10] The ICAO standards and recommended practices TSA assesses foreign airports against are referred to collectively in this report as ICAO standards or standards. See appendix III for a description of the ICAO standards TSA uses to assess security measures at foreign airports.[11]

TSA uses a risk-informed approach to schedule foreign airport assessments by categorizing airports into three tiers.[12] Specifically, Tier 1 airports—airports that are determined to be low risk—are assessed once every 3 years; Tier 2 airports—airports determined to be medium risk— are assessed every 2 years; and Tier 3 airports—those determined to be high risk—are assessed annually. TSA's assessments of foreign airports are conducted by a team of inspectors, which generally includes one team leader and one team member. According to TSA, it generally takes 3 to 7 days to complete a foreign airport assessment. However, the amount of time required to conduct an assessment varies based on several factors, including the size of the airport, the number of air carrier station inspections to be conducted at the airport, the threat level to civil aviation in the host country, and the amount of time it takes inspectors to travel to and from the airport where the assessment will take place.[13]

TSA uses a multistep process to conduct assessments of foreign airports. Specifically, the TSA Representative (TSAR) must obtain approval from the host government to allow TSA to conduct an airport assessment, and schedule the date for the on-site assessment. After conducting an entry briefing with Department of State, host country officials, and airport officials, the team conducts an on-site visit to the airport. During the assessment, the team of inspectors uses several methods to determine a foreign airport's level of compliance with ICAO standards, including conducting interviews with airport officials, examining documents pertaining to the airport's security measures, and conducting a physical inspection of the airport. For example, inspectors are to examine the integrity of fences, lighting, and locks by walking the grounds of the airport. Inspectors also make observations on access control procedures, such as looking at employee and vehicle identification methods in secure areas, as well as monitoring passenger and baggage screening procedures in the airport. At the close of an airport assessment, inspectors brief foreign airport and government officials on the results of the assessment. TSA inspectors also prepare a report summarizing their findings on the airport's overall security posture and security measures, which may contain recommendations for corrective action and must be reviewed by the TSAR, the ROC manager, and TSA headquarters officials. See appendix IV for more information on the multistep process TSA uses to conduct its assessments of foreign airports.

Along with conducting airport assessments, the same TSA inspection team also conducts air carrier inspections when visiting a foreign airport to ensure that air carriers are in compliance with TSA security requirements. Both U.S. air carriers and foreign air carriers with service to the United States are subject to inspection. When conducting air carrier inspections, TSA inspectors examine compliance with applicable security requirements, including TSA-approved security programs, emergency amendments to the security programs, and security directives.[14] As in the case of airport assessments, air carrier inspections are conducted by a team of inspectors, which generally includes one team leader and one team member. An inspection of an air carrier typically takes 1 or 2 days, but can take longer depending on the extent of service by the air carrier. Inspection teams may spend several days at a foreign airport inspecting air carriers if there are multiple airlines serving the United States from that location. During an inspection, inspectors are to review applicable security manuals, procedures, and records; interview air carrier station personnel; and observe air carrier employees processing passengers from at least one flight from passenger check-in until the flight departs the gate to ensure that the air carrier is in compliance with applicable requirements. Inspectors evaluate a variety of security measures, such as passenger processing, checked baggage acceptance and control, aircraft security, and passenger screening. If an inspector finds that an air carrier is not complying with applicable security requirements, additional steps are to be taken to record such instances and, in some cases, pursue them with further investigation.

If the inspectors report that an airport's security measures do not meet minimum ICAO standards, particularly critical standards, such as those related to passenger and checked baggage screening and access controls, TSA headquarters officials are to inform the Secretary of Homeland Security.[15] If the Secretary, based on TSA's airport assessment results, determines that a foreign airport does not maintain and carry out effective security measures, he or she must, after advising the Secretary of State, take secretarial action. See appendix II for a detailed description of the process for taking secretarial actions against a foreign airport.

GAO's 2007 Review of TSA Foreign Airport Assessment Program

In 2007, we issued a report on TSA 's foreign airport assessment program, including the results of TSA's foreign airport assessments, actions taken and assistance provided by TSA when security deficiencies were identified at foreign airports, TSA oversight of its program, and TSA's efforts to address challenges in conducting foreign airport assessments. Specifically, we reported that TSA's oversight of the foreign airport assessment program could be strengthened. For example, TSA did not have adequate controls in place to track whether scheduled assessments and inspections were actually conducted, deferred, or canceled. TSA also did not always document foreign officials' progress in addressing security deficiencies identified by TSA. Further, TSA did not have outcome-based performance measures to assess the impact of its assessments on the security of U.S.-bound flights. As a result, we recommended that TSA develop controls for tracking and documenting information and establish outcome-based performance measures to strengthen oversight of its foreign airport and air carrier evaluation programs. DHS concurred with the recommendations and has since taken several actions to address them, which we discuss later in our report.

TSA HAS TAKEN STEPS TO ENHANCE FOREIGN AIRPORT ASSESSMENTS, BUT CHALLENGES REMAIN

TSA Steps to Update and Streamline Its Assessments

Since 2007, TSA has taken a number of steps to update and streamline its foreign airport assessment program, as discussed below.

TSA revised and updated its Standard Operating Procedures (SOP) for the program. In 2010, TSA revised the SOP, which prescribes program and operational guidance for assessing security measures at foreign airports.

TSA also streamlined the assessment process by reducing the number of ICAO standards it assesses foreign airports against from 86 to 40.[16] Of the 40, TSA officials we interviewed told us the agency has identified 22 standards as key for determining an airport's level of security.[17] In addition, TSA reduced the assessment report writing cycle time for inspectors from 38 calendar days to 20 calendar days, which was intended to expedite the delivery of assessment reports to host governments. This new requirement has helped TSA reduce the time needed to deliver its assessment results to foreign countries, but all 23 inspectors we interviewed told us this requirement was often difficult to meet due to a variety of factors. For example, upon returning from a visit, TSA inspectors reported that they need to document both the airport assessment and air carrier inspections, and plan their next trip, which makes the reduced reporting time requirement difficult to meet. However, the Director of Global Compliance told us that for larger airports with many air carriers, TSA recently began separating the airport assessment and air carrier inspection visits into two separate visits, thus reducing the documentation workload. Moreover, the deadline to submit documentation has been delayed for some back-to-back assessment trips in order to provide sufficient time for inspectors to complete the documentation.[18] The Director of Global Compliance also stated that, in fiscal year 2012, all employees will have training opportunities in order to improve writing skills and reduce the amount of time dedicated to editing and rewriting assessments. In addition, to address resource needs we identified in 2007, TSA hired 6 additional international inspectors in 2007 and 10 international cargo inspectors in 2008 and created 25 new international inspector positions, of which 15 were filled as of July 2011.[19] TSA plans to fill the remaining 10 positions by the end of 2011.[20] The Director of Global Compliance stated that the burden of writing and processing assessment reports should be lessened as the agency hires additional inspectors because this will create a greater pool of available inspectors to conduct and document the assessments.

TSA implemented a new risk-informed methodology for prioritizing and scheduling its assessments at foreign airports in 2010.[21] Specifically, TSA now categorizes foreign airports as high, medium or low risk. Of the roughly 300 foreign airports TSA assesses, TSA identified some airports as high risk and others as medium risk as of August, 2011.[22] The remaining airports were deemed low risk.[23] TSA's methodology for determining an airport's risk category is based on the likelihood of a location being targeted (threat), the protective measures in place at that location (vulnerability), and the potential impact of an attack on the international transportation system (consequence). TSA uses current threat information, airport passenger and flight data, and prior airport assessment results to assign each airport a numerical risk score, which is then used to determine its overall risk ranking. As part of this calculation, TSA assigns each airport an overall vulnerability score of 1–5. These scores, or categories, are numerical representations of compliance or noncompliance with the ICAO standards the agency assesses each foreign airport against. Specifically, using an airport's most recent assessment report, the ROC Manager and TSA's Director of Global Compliance assign an overall vulnerability category for each airport based on the following descriptions provided in the 2010 *Foreign Airport Assessment Program SOP*:

- Category 1: Fully Compliant;
- Category 2: Capability Exists with Minor Episodes of Noncompliance;
- Category 3: Capability Exists, Compliance is Generally Noted, Shortfalls Remain;
- Category 4: Capability Exists, Serious Lack of Implementation Observed; and
- Category 5: Egregious Noncompliance.

Once the vulnerability score is determined, it is then combined with each airport's related threat and consequence information to determine its risk category. TSA attempts to assess high-risk airports every year, medium- risk airports once every 2 years, and low-risk airports once every 3 years. TSA's Director of Global Compliance told us this new approach allows the agency to better allocate resources to identify and mitigate security concerns at foreign airports it assesses. In addition, all the TSA ROC managers and 19 of the 23 inspectors we interviewed during our site visits told us that this new foreign airport risk prioritization methodology was an improvement over the previous process.[24] These officials also stated that this new approach has helped them reduce the number of assessments conducted annually, enabling inspectors to better adhere to the annual schedule. On the basis of our analysis, TSA's approach for scheduling foreign airport assessments is consistent with generally accepted risk management principles, which define risk as a function of threat, vulnerability, and consequence.[25]

TSA developed a 2011 strategic implementation plan. This plan establishes annual program objectives and milestones, and links program activities to broader agency aviation security goals providing a road map for their completion.[26]

TSA began declassifying its foreign airport assessment reports. Since 2007, TSA has been declassifying the reports from Confidential and designating them SSI to facilitate better access to and the dissemination of program results, while still providing protection for foreign government information deemed sensitive. TSA officials noted that the declassification of assessment results is essential for TSA because staff could not easily access the specifics of prior results and deficiencies from reports that have not yet been declassified.

TSA formed the Capacity Development Branch (CDB). TSA created the CDB in 2007 to manage all TSA international aviation security capacity building assistance efforts, including requests for assistance in response to a host government's airport assessment results. Through CDB, TSA provides six aviation security training courses that address, among other things, preventive security measures, incident management and response, and cargo security.[27] In 2008, TSA also developed the Aviation Security Sustainable International Standards Team (ASSIST) Program to provide more long-term, sustainable, technical aviation security assistance to select foreign countries. Thus far, TSA has partnered with five countries under the ASSIST program: St. Lucia, Liberia, Georgia, Haiti, and Palau. See appendix V for more specific information on TSA assistance provided these countries under ASSIST.

TSA developed assessment tracking tools to provide better oversight of program information. In 2007 we reported that TSA did not have controls in place to track the status of scheduled foreign airport assessments, including whether assessments were actually conducted or whether they were deferred or canceled, which could make it difficult for the agency to ensure that scheduled assessments are actually completed. We also reported that TSA did not always document the results of follow-up conducted by TSA staff to determine progress made by foreign governments in addressing security deficiencies identified by TSA inspectors during assessments, and that such follow-up would enable the agency to have access to updated information on the security status of foreign airports that provide service to the United States. In response to the findings and recommendations we made in our 2007 report, TSA implemented a tool to track its annual foreign airport assessment schedule, including reasons why assessments were deferred or canceled, and a tracking sheet to compile the results of its prior airport assessments. Specifically, this sheet documents the frequency with which foreign airports complied with particular categories of ICAO standards, such as passenger screening, baggage screening, and access controls, among others. TSA also developed a tool whereby deficiencies previously identified during an assessment can be tracked to monitor the progress made by host governments in rectifying security deficiencies. TSA's Director of Global Compliance told us these tracking sheets have helped TSA provide better oversight and monitoring of key program information.

TSA signed several working arrangements to facilitate its assessments. Since 2007, TSA has signed a multilateral working arrangement with the European Union (EU), and several bilateral working arrangements with individual foreign nations, to facilitate, among other things, TSA assessments at foreign airports. Specifically, in 2007, we reported that TSA had taken steps toward harmonizing[28] airport assessment processes with the European Commission (EC).[29] As part of these efforts, TSA and the EC established six working groups to facilitate, among other things, sharing of SSI between TSA and the EC, TSA observation of EU airport assessments, as well as EC observation of TSA assessments of airports in the United States.[30] In 2008, TSA signed a multilateral working arrangement with the EU to facilitate joint assessments and information sharing between TSA and the EU. Specifically, under the arrangement, TSA and the EC coordinate assessment schedules annually to identify airport locations at which to conduct joint assessments. EC officials we interviewed told us their main goal under the arrangement was to better leverage resources and reduce the number of TSA visits per year to European airports because of concerns from EU member states on the frequency of visits from EC and U.S. audit teams. TSA officials we interviewed said they also wanted to better leverage existing resources while ensuring continued TSA access to European airports for the purposes of conducting security assessments. While TSA agreed to conduct assessments at EU airports no more than once every 5 years, EU and TSA officials we interviewed said the EC permits TSA to approach a country bilaterally if scheduling conflicts do not allow for an assessment to be conducted jointly. TSA also occasionally conducts table-top reviews in place of on-site airport visits. Specifically, if the EC inspected an airport within the last 2 years, TSA will sometimes meet with EC officials to review the EC inspection report—referred to as a table top—which typically contains enough information for

TSA to make its evaluations.[31] However, TSA officials said table-top reviews should not serve as a permanent substitute for TSA onsite assessments.

TSA has also entered into several bilateral working arrangements with foreign countries to facilitate its airport assessments. Specifically, TSA has signed arrangements with Brazil, Germany, India, the United Kingdom, Russia, and is in the process of establishing arrangements with Nicaragua and Portugal. These arrangements specify certain conditions, practices, and protocols for sharing key information with TSA, but also impose some constraints, such as limiting the number of TSA visits per year, including the length of the visit.

Challenges Affecting TSA Assessment Efforts

Even with TSA's efforts to enhance the program, challenges remain in several areas: gaining access to some foreign airports, developing an automated database to manage program information, prioritizing and providing training and technical assistance, and expanding the scope of TSA's airport assessments to include all-cargo operations, as discussed below.

TSA access to some foreign airports has been limited by sovereignty concerns. In 2007, we reported that some host governments expressed concerns that TSA assessments infringe upon their authority to regulate airports and air carriers within their borders, and that some foreign governments had denied TSA access to their airports. TSA's multilateral and bilateral arrangements have helped to facilitate assessments in some foreign countries, but TSA has had difficulty gaining access to some foreign airports due to sovereignty concerns raised by host governments. For example, TSA has not been able to assess any of the four airports in Venezuela or conduct TSA compliance inspections for air carriers flying out of Venezuela into the United States, including U.S. air carriers, since 2006. Thus, TSA has been unable to determine the security posture of flights from Venezuela bound for the United States. On September 8, 2008, the Secretary of Homeland Security issued a Public Notice that informs the public that the U.S. Government is unable to determine whether airports in Venezuela that serve as the last point of departure for nonstop flights to the United States maintain and carry out effective aviation security measures.[32]

A TSA official told us that a TSA representative traveled to Venezuela recently to start discussions with the Venezuelan government about TSA regaining access to Venezuelan airports to conduct assessments and air carrier inspections. Since it is unclear what the outcome of these discussions will be, and when TSA will regain access to airports and air carriers in Venezuela, the Public Notice remains in effect. Until TSA is able to regain access to airports and air carriers in Venezuela to conduct assessments and air carrier inspections, the agency will be unable to determine to what extent, if at all, airports in Venezuela are maintaining and carrying out effective security measures, or the extent to which air carriers are complying with TSA security requirements for U.S.-bound flights.[33] The Director of Global Compliance indicated TSA is concerned about sovereignty issues with other foreign countries and their willingness to allow TSA inspectors to assess their airports and air carriers. TSA has been working on establishing a Memorandum of Understanding with one country to ensure continued TSA access to its airports. Moreover, TSA indicated that working arrangements it developed with two other countries were undertaken to address government sovereignty concerns over TSA's assessments.

TSA has experienced difficulties developing an automated database. Since 2007, TSA has been in the process of trying to develop an integrated, automated database management system to allow for more timely submission of foreign airport assessment results, as well as perform more substantive analysis and comparisons of foreign airport trends and issues. Specifically, in response to our 2007 recommendations, TSA stated that they were exploring an automated means of capturing foreign airport assessment data to track airport deficiencies identified, corrective actions recommended by TSA, and any resulting actions taken by the host nation. In 2010, TSA field tested a system, called the Foreign Airport Assessment Reporting System (FAARS), which was intended to store results of airport assessments for easier data extraction and manipulation. For example, while airport assessments are currently prepared as word documents (typically around 60 pages in length), FAARS was intended to put information into database fields, which would have allowed the Office of Global Strategies (OGS) to run reports on specific indicators, such as which foreign airport checkpoints are using Advanced Imaging Technology (AIT) units.[34] However, the Director of Global Compliance told us FAARS ultimately did not meet TSA's needs and was discontinued because, among other things, data entry was cumbersome and certain data fields could not be edited. Further, the database was not web-based, and instead had to be installed on users' hard drives, not allowing for easy integration of multiple users and data.

In April 2011, TSA developed a comprehensive functional requirements document, which outlines the capabilities and functions required for a new proposed software solution. TSA officials told us they provided it to officials in TSA's Offices of Acquisition and Information Technology who developed a contract for developing, testing, fielding, and distributing a software solution that meets programs needs. TSA officials told us that the contractor who will develop the product has received the Statement of Work, and initial implementation of the product is planned for fiscal year 2012, with full capability planned to follow in fiscal year 2013. Given these time frames, it will be important for TSA to monitor the status of this effort to ensure a solution is implemented within reasonable time frames, particularly since we raised this issue in our 2007 report and it is still not clear when a solution will be fully vetted and implemented. TSA's Director of Global Compliance also told us that identifying a database management system that meets of the needs of the program has been a long-standing challenge for the program.

TSA's training and technical assistance efforts face several challenges, and TSA's new equipment loan program has raised concerns. TSA has initiated several capacity building efforts since our 2007 report, but these efforts have been affected by conflicting Department of State priorities, and TSA's new equipment loan program has raised concerns about ensuring that loaned equipment is properly operated and maintained. Specifically, in addition to its own training courses and technical assistance, CDB provides training and technical assistance sponsored by the Department of State's Anti-Terrorism Assistance (ATA) program[35] and from the Organization of American States Inter-American Committee Against Terrorism,[36] which is funded through the State Department. A CDB official stated they currently have eight employees and limited funds to provide aviation security technical and training assistance to partner nations overseas.[37] As a result, a CDB official told us their training schedule often has a 3-month lag from when training is requested to when it is provided. In addition, four TSARs we spoke with stated they sometimes have difficulty getting their requests for TSA training from host nations fulfilled because of a lack of resources. According to a TSA official we spoke with, during the past 2 years, the U.S. government's aviation security training and assistance priorities have been largely driven by State Department priorities. For example, of the 64 course offerings CDB had planned to provide in 47 foreign countries at the beginning of fiscal year 2011, 33 were sponsored by State ATA or the Organization of American States, and some number of those countries have high-risk airports as identified by TSA. In addition, TSA's 2010 training schedule showed that of the 53 course offerings CDB provided in 33 countries, 29 were sponsored by State ATA or the Organization of American States, and some number of those countries have high-risk airports as identified by TSA.[38] CDB and State Department officials told us they plan to work more closely in the future to better align their respective priorities.

In addition to providing various types of training and technical assistance, TSA has also provided aviation security equipment to foreign countries to help these countries enhance their existing capabilities and practices. Specifically, one of TSA's goals in its CDB fiscal year 2011–2015 Strategic Plan is to develop the necessary procedures for a system of long-term lending of decommissioned TSA screening equipment to partner countries.[39] In accordance with authority granted under the Aviation and Transportation Security Act, TSA has undertaken to provide or loan security technologies and other equipment to foreign governments.[40] According to TSA officials, the agency exercises this authority in coordination with the Department of State, and has obtained authority from the Department of State to negotiate and conclude agreements with foreign governments to provide technical cooperation and assistance, referred to as "Circular 175" agreements.[41] For example, following the October 2010 discovery of explosive devices in air cargo packages bound for the United States from Yemen, TSA loaned six hand-held explosives trace detection devices to Yemen in an expedited fashion as a response to an emergent threat to help enhance the government's passenger and cargo screening processes.[42] TSA officials also told us that the agency has provided security technology and equipment to Aruba, Bahamas, Bermuda, Haiti, Ireland, and Malta under this same authority.

While TSA has provided some equipment to foreign countries, TSA and EC officials we spoke with identified potential challenges associated with doing so. For example, TSA officials cited some foreign governments' inability to properly maintain and operate TSA-provided screening equipment once provided. TSA officials told us it will be important for the agency to ensure that a foreign government has the appropriate staff, and that they are properly trained and ready to operate the equipment as well as conduct any necessary maintenance, to ensure that the U.S.-provided equipment is being used as intended and remains operational. TSA officials also explained that while under its existing authority it can donate or otherwise transfer equipment, such authority does not authorize TSA to provide maintenance and service contracts for this equipment. TSA officials we spoke with told us they would support congressional efforts to provide the agency with this additional authority. In addition, EC officials we interviewed identified similar challenges to their current and potential future efforts to provide various types of capacity building assistance to foreign countries. TSA officials said it will be important for TSA to establish user agreements with recipient countries that ensure U.S. government resources are not wasted or inappropriately used.

Several factors may complicate TSA assessments of foreign all-cargo operations. Following the attempted bombing of an all-cargo flight bound for the United States from Yemen in October 2010, TSA decided to devote additional resources to assessing all-cargo airports. While TSA is still in the early planning stages of its efforts to assess all-cargo operations at foreign airports, several factors may complicate these efforts.[43] Specifically, TSA's Director of Global Compliance stated that the agency has identified 17 foreign airports that serve as all-cargo last points of departure to the United States. As of July 2011, TSA has conducted two all-cargo assessments of two airports in China. Moreover, TSA plans to assess two additional all-cargo airports by the end of fiscal year 2011. According to TSA, from these first visits, TSA is making some adjustments to the assessment process. For fiscal years 2012 through 2013, TSA plans to schedule visits to the remaining 15 airports that serve as all-cargo last points of departure to the United States, pending host government permission. However, TSA stated that it is too early to tell how many additional inspectors may be needed to complete these assessments.

TSA officials we interviewed identified several factors that may complicate TSA's assessments of all-cargo operations at foreign airports. For example, all of the 23 TSA inspectors we interviewed expressed concerns about incorporating additional assessment visits into their annual schedules given their current workloads. In addition, these officials stated that it is uncertain whether foreign governments will allow TSA inspectors to assess their all-cargo operations and all-cargo airports. For example, while TSA has several bilateral arrangements with foreign countries to facilitate its assessments, TSA officials told us these arrangements do not specify access to cargo operations or all-cargo airports. Moreover, all four cargo inspectors we met with said it is logistically difficult to assess "upstream" cargo originating from other non-last point of departure airports. These inspectors said these logistical challenges will be an important factor for the agency to consider when selecting foreign airports to assess as well as in making determinations on the security posture of cargo on flights departing foreign airports for the United States. In addition, these inspectors also said that travel to some foreign all-cargo operation airports may be logistically difficult because of the lack of direct passenger flights and may require long travel by car or train. The Director of Global Compliance acknowledged that this new effort is challenging and stated that the agency will address these issues on a case-by-case basis. However, the Director also stated that with the increase to the inspector workforce, the cross-training of generalist international aviation inspectors to perform cargo inspections, and the limited additional locations to visit, TSA will be able to perform these additional visits over the next 2 years.

SOME NUMBER OF FOREIGN AIRPORTS COMPLIED WITH ICAO STANDARDS, BUT TSA COULD BETTER USE ITS ASSESSMENT RESULTS

Some Foreign Airports Complied with ICAO Standards, but TSA Identified Serious Noncompliance Issues at Other Airports

Based on our analysis of the results of TSA's foreign airport assessments conducted during fiscal year 2006 through May 9, 2011, some number of the foreign airports TSA assessed complied with all of TSA's aviation security assessment standards.[44] However, TSA has identified serious or egregious noncompliance issues at a number of other foreign airports. Common areas of noncompliance included weaknesses in airport access controls and passenger and baggage screening. Moreover, our analysis of TSA's assessments showed variation in compliance across regions, among various individual standards, and by airports' risk level. For example, our analysis showed that some number of regions of the world had no airports with egregious noncompliance while some regions had several such airports. Specific information related to our analysis of TSA's airport assessment results is deemed SSI.

TSA Has Not yet Analyzed Its Assessment Results to Identify Trends and Better Inform Future Activities

TSA has not taken steps to analyze or evaluate its foreign airport assessment results in the aggregate to identify regional and other trends over time, which could assist the agency in informing and prioritizing its future activities. TSA officials have access to results of foreign airport assessments dating back to fiscal year 1997, but they have not analyzed the information to gain insight into how foreign airports' security posture may have changed over time or identified regional and other patterns and trends over time. Specifically, TSA's airport assessment reports are collected in an online repository that can be accessed by employees, and TSA's Director of Global Compliance compiles high-level information from each airport assessment in a tracking tool, which allows her to view the overall results of assessments without having to go back to individual narrative reports.[45] However, according to TSA, the agency has not analyzed the data contained in this tracking tool, which could assist TSA in informing and prioritizing its future activities and assessing the results of its past assessment efforts. In addition, while the spreadsheet provides a snapshot of airports and their results compared to the ICAO standards, it does not indicate why a standard was not met by an airport. If TSA employees would like to know why a certain airport did not meet a standard in a previous year, they must locate and read the report for that assessment. TSA's Director of Global Compliance told us that this is labor intensive, and makes it difficult to identify anomalies or trends over time.

Standards for Internal Control in the Federal Government require agencies to ensure that ongoing monitoring occurs during the course of normal operations to help evaluate program effectiveness.[46] TSA's Director of Global Compliance as well as all TSA ROC managers and inspectors we interviewed agreed that information pertaining to identified vulnerabilities in foreign airports should be compiled in regional-, country-, and airport-specific aggregates to help conduct planning and assess the results of program activities. TSA's Director of Global Compliance stated that TSA has prepared a vacancy announcement for a program analyst position which may, when filled, be tasked with compiling overall results and analyzing assessment results.

TSA's Director of Global Compliance as well as all ROC managers and inspectors we interviewed also agreed that analysis of foreign airport assessment results would be helpful in identifying the aviation security training needs of foreign aviation security officials. TSA has one internally funded program in place that is specifically intended to provide aviation security training and technical assistance to foreign aviation security officials. However, TSA also coordinates with other federal agencies, such as the Department of State, to identify global and regional training needs and provide instructors for the aviation security training courses State offers to foreign officials. While TSA does not always determine which foreign countries receive aviation security training and technical assistance offered by other federal agencies, TSA could use the cumulative results of TSA's foreign airport assessments to better support TSA's priorities for aviation security training and technical assistance. Moreover, with analysis of airport assessment results, TSA could better inform its risk management decision making by identifying trends and security gaps, and target capacity building efforts accordingly. Specifically, this evaluation could include an analysis of the frequency of noncompliance issues TSA inspectors identified, including regional variations and perspectives on the security posture of individual airports over time. Further, a mechanism to evaluate cumulative foreign airport assessment results could help the agency better allocate and target its future resources and better understand its results, including the impact the program is having on enhancing foreign nations' ability to comply with ICAO standards.

TSA Has Not yet Developed Outcome-Based Performance Measures

In 2007, we reported that TSA was taking steps to assess whether the goals of the foreign airport assessment program were being met, but that it had not yet developed outcome-based performance measures to evaluate the impact TSA assistance has on improving foreign airport compliance with ICAO standards. As a result, we recommended that TSA establish outcome-based performance measures to strengthen oversight of the program.[47] While DHS officials agreed with the recommendation in 2007, according to TSA, the agency has not yet developed such measures. The goal of the foreign airport assessment program is to ensure the security of U.S.-bound flights by evaluating the extent to which foreign governments are complying with applicable security requirements. The Government Performance and Results Act (GPRA) of 1993, as amended by the GPRA Modernization Act of 2010, requires executive branch departments to use performance measures to assess progress toward meeting program goals and to help decision makers assess program accomplishments and improve program performance.[48] Performance measures can be categorized either as outcome measures, which describe the results of carrying out a program or activity, or as output measures, which describe the direct products or services delivered by a program or activity, or as process measures, which address the type or level of program activities conducted, such as timeliness or quality. TSA has taken some steps to develop a variety of measures and is reporting this information to the Office of Management and Budget. These measures include:

- average number of international inspections conducted annually per inspector,
- percentage of foreign airports serving as Last Point of Departure operating in compliance with leading security indicators,[49]
- percentage of countries with direct flights to the U.S. that are provided aviation security assistance, and
- percentage of countries/territories with no direct flights to the U.S. that are provided aviation security assistance.

While these measures are useful in determining, for example, the percentage of airports operating in compliance with security indicators, they do not address the ultimate results of the program, as outcome measures could. Outcome-based measures could help determine the extent to which TSA programs that assess and provide training and technical assistance to foreign airports have helped to improve security at airports that service the United States. However, TSA's Director of Global Compliance noted several possible challenges with applying such outcome measures to the assessment program. Specifically, the Director stated that the foreign airport assessment program is designed to identify—not correct—security deficiencies at foreign airports, and that whether or not foreign officials improve security at their airports is not within TSA's control. The Director added that such measures may create a disincentive for inspectors to objectively assess an airport's level of compliance. Despite these challenges, the Director acknowledged the importance of developing outcome measures and stated that their development should be the responsibility of TSA's Office of Global Strategies, not individual programs within this office, such as the foreign airport assessment program that she leads. Even without full control over the outcomes associated with such measures, we continue to believe our prior recommendation is still valid and that it would be useful for TSA to develop reasonable outcome-based measures, such as the percentage of security deficiencies that were addressed as a result of TSA onsite assistance or related technical assistance and training offered by the CDB, and TSA recommendations for corrective action. As we previously recommended, such measures would help TSA establish greater accountability over the way in which TSA uses its resources and, in conjunction with its existing measures, enable the agency to evaluate and improve the impact of its assistance on improving security at foreign airports.

OPPORTUNITIES EXIST TO CLARIFY AIRPORT ASSESSMENT CRITERIA, FURTHER TARGET AIRPORT ASSESSMENTS, AND SYSTEMATICALLY IDENTIFY SECURITY BEST PRACTICES

While TSA has taken a number of steps to improve and streamline its foreign airport assessment program since our 2007 report, opportunities exist for TSA to make additional improvements in several key areas. For example, TSA has taken steps to make its foreign airport assessments more risk informed, but the agency lacks clearly defined criteria to determine a foreign airport's level of noncompliance with ICAO standards. For example, as stated earlier, TSA provides each airport an overall vulnerability category, or score, of 1 through 5, which is a numerical representation of compliance or level of noncompliance with the ICAO standards the agency assesses each foreign airport against. However, TSA has not developed any specific criteria, definitions, or implementing guidelines to ensure ROC managers and other program management officials apply these categories consistently across airports. For example, the SOP does not define how to assess whether an airport should receive a vulnerability rating of 3—"capability exists, compliance is generally noted, shortfalls remain," versus a vulnerability rating of 2—"capability exists with minor episodes of noncompliance." In the absence of more specific and transparent criteria and guidance, it is not clear how TSA applied these related categories—which describe the level of noncompliance—to the results of the assessments, or whether they were applied consistently over time. The lack of documented guidance prevented us from making an analysis or comparison of how TSA made its determinations. This is particularly important given that these scores represent an overall assessment of an airport's level of compliance or noncompliance with ICAO standards that TSA has deemed critical to airport security, and also are a key component of TSA foreign airport risk- ranking determinations.

TSA's Director of Global Compliance agreed these category determinations are largely subjective judgments based on many facts and circumstances. TSA's Director of Global Compliance stated that it is challenging to establish specific guidance for how to assign these categories because of the numerous factors that can influence the decision for assigning vulnerability scores. The Director also noted that because she reviews each assessment report and weighs in on each assigned category, she in effect serves to institutionalize the scores and ensure they are consistent from airport to airport. *Standards for Internal Control in the Federal Government* call for controls and other significant events to be clearly documented in directives, policies, or manuals to help ensure operations are carried out as intended.[50] This is especially important should key staff leave the agency. Although we recognize the inherently subjective nature of the standards, providing TSA decision makers with more specific criteria and definitions for determining a foreign airport's level of compliance with ICAO standards would provide greater assurance that such determinations are consistent across airports over time. The Director acknowledged that additional guidance, such as examples to illustrate what these categories mean, could help ensure greater transparency and consistency over how airport vulnerability scores are determined. Such consistency is important since airport vulnerability determinations are used to calculate an airport's overall security risk level, which in turn affects the program's activities and resource needs.

In addition, TSA officials we spoke with identified opportunities for TSA to increase program efficiency by conducting more targeted airport assessments. Specifically, ROC managers and inspectors at all the locations we visited stated there are opportunities for TSA to conduct more targeted, smaller scale assessments at foreign airports that could focus more exclusively on the key security issues at a particular airport rather than having inspectors conduct full-scale assessments every visit. For example, the ROC Manager of one location we visited stated that the Federal Aviation Administration previously conducted supplemental-type visits of foreign airports that were reduced in scope and only focused on specific issues or deficiencies that needed to be addressed. He said that TSA should consider ways to incorporate this type of assessment philosophy into its current operations as it may help further streamline the assessment process and associated time frames. ROC managers at all the locations we visited also said inspectors often know, from their prior visits and assessment reports, what specific issues are present at specific airports, and that focusing more time on key issues could provide a more effective way of addressing and correcting security deficiencies. Twenty of 23 inspectors we spoke with said this type of assessment would also reduce repetitive and duplicative data gathering. In addition, these inspectors stated they sometimes do not have the opportunity to conduct all necessary onsite operational observations, document reviews, and interviews because they spend a significant amount of time addressing other descriptive, less critical aspects of the assessment. They said more targeted risk-informed assessments would allow them to focus more time and attention on key security issues, resulting in higher quality and more useful assessment results. Exploring opportunities to conduct more targeted assessments could help TSA enhance the efficiency and value of TSA's foreign airport assessment program.

TSA's Director of Global Compliance told us they have begun to conduct abbreviated and targeted airport assessments in some cases due to the security risks associated with traveling and working in certain countries. For example, in 2011 TSA conducted abbreviated assessments at airports in Mexico and Iraq, due to the current security situation, which focused on a select number of critical areas rather than on all topics typically covered during an assessment. While targeted or abbreviated assessments are viewed as beneficial in some circumstances, TSA's Director of Global Compliance also stated that conducting a comprehensive assessment is important because inspectors may visit an airport only once every 3 years, to document any security changes, deficiencies, or improvements since the previous visit. The Director also raised a concern about conducting additional targeted assessments if they limited opportunities to conduct regularly scheduled comprehensive assessment visits. However, we believe TSA's use of abbreviated or targeted assessments could be expanded in cases where it would not have a negative impact on the program. For example, as TSA works to systematically analyze the results of its assessments, it may determine that specific regions of the world need additional assistance in meeting certain critical standards. TSA could use this information to focus or target its assessments to address these higher risk scenarios, thus leveraging program resources. Such efforts are consistent with TSA's ongoing risk-informed activities, as discussed earlier in this report. Moreover, we have previously reported that risk-informed, priority driven decisions can help inform decision makers when allocating finite resources to the areas of greatest need.[51] In addition, TSA has not taken steps to systematically compile or analyze security best practice information that could contribute to enhancing the security of both foreign and U.S. airports. TSA officials acknowledged possible opportunities to better identify, compile, and analyze aviation security best practices through their assessments at foreign airports. We have previously reported that in order to identify innovative security practices that could help further mitigate terrorism-related risk to transportation sector assets, it is important to assess the feasibility as well as the costs and benefits of implementing security practices currently used by foreign countries.[52] While TSA compiles information in its foreign airport assessment reports to evaluate the degree to which airports are in compliance with select ICAO standards, it does not have a process in place to identify and analyze aviation security best practices that are being used by foreign airports to secure their operations and facilities. TSA officials agreed that identifying relevant best practices could help TSA better leverage their assessment activities by assisting foreign airports in increasing their level of compliance with ICAO standards, as well as in identifying security practices and technologies that may be applicable to enhancing the security of U.S. airports.

In December 2, 2010, testimony before the Senate Committee on Commerce, Science and Transportation, TSA's Director of Global Compliance confirmed that there are a variety of ways in which foreign airports can effectively meet ICAO standards. For example, one airport might address access control security by using coded door locks and swipe cards, while another may lock its doors and limit the number of available keys to certain personnel. Airports may also establish perimeter security in different ways, such as through fencing or natural barriers. In addition, TSA inspectors, as part of the assessment, often obtain detailed information and understanding of the various types of security technologies and methods being used by foreign governments, which may also be applicable and cost-effective for U.S. airports. For example, while accompanying TSA inspectors during an airport assessment, we observed TSA inspectors being briefed on various passenger screening processes, technologies, and equipment that were comparable to, and in some cases may have exceeded, those used in the U.S. We believe establishing a mechanism to systematically compile and analyze this type of information could help ensure TSA is more effectively able to assist foreign airports in meeting ICAO standards and improve security practices, as well as identify security practices and technologies that may be applicable to enhancing the security of U.S. airports.

CONCLUSION

Securing commercial aviation operations remains a daunting task—with hundreds of airports and thousands of flights carrying millions of passengers and pieces of checked baggage to the United States every year. TSA's foreign airport assessment program is aimed at enhancing this system by identifying critical security weaknesses and gaps in airports serving the United States, which in turn can help inform and guide needed efforts to mitigate these deficiencies. TSA has taken a number of actions to enhance its foreign airport assessment program since 2007, but additional steps can help further strengthen the program.

For example, developing a mechanism to evaluate assessment results to determine security trends and patterns could enable TSA to target and prioritize future assessment activities, including training and other capacity building resources.

Moreover, establishing criteria and guidance for determining the vulnerability of individual foreign airports would provide for greater consistency of these vulnerability ratings across airports over time. Such consistency is important since airport vulnerability determinations are used to calculate an airport's overall security risk level. Further, exploring the feasibility of conducting more targeted assessments could help enhance the efficiency and value of TSA's foreign airport assessment program. Moreover, systematically compiling information on aviation security best practices could help ensure TSA is more effectively able to assist foreign airports in meeting ICAO standards and improve security practices, as well as identifying security practices and technologies that may be applicable to enhancing the security of U.S. airports.

RECOMMENDATIONS FOR EXECUTIVE ACTION

To help further enhance TSA's foreign airport assessment program, we recommend that the Secretary of Homeland Security direct the Assistant Secretary for the Transportation Security Administration to take the following three actions:

- Develop a mechanism to evaluate the results of completed assessment activities to determine any trends and target future activities and resources. This evaluation could include frequency of noncompliance issues, regional variations, and perspectives on the security posture of individual airports over time.
- Establish criteria and guidance to assist TSA decision makers when determining the vulnerability rating of individual foreign airports.
- Consider the feasibility of conducting more targeted assessments and systematically compiling information on aviation security best practices.

AGENCY COMMENTS AND OUR EVALUATION

We provided a draft of the sensitive version of this report to DHS and TSA on September 1, 2011, for review and comment. In commenting on our report, DHS stated that it concurred with all three of the recommendations and identified actions taken or planned to implement them. DHS also highlighted new initiatives under way by the Office of Global Strategies. Regarding the first recommendation that TSA develop a mechanism to evaluate the results of completed assessment activities to determine any trends and target activities and resources, and that this evaluation could include frequency of noncompliance issues, regional variations, and perspectives on the security posture of individual airports over time, DHS concurred. DHS stated that TSA has taken several steps to address this recommendation including utilizing a program analyst to create analyses reflecting temporal and site-specific trends and anomalies. DHS also stated that TSA established a project team to evaluate regional, country, and airport vulnerabilities and determine those problem areas that could be effectively addressed by training. DHS also noted that TSA is developing workshops that can be presented by inspectors at the conclusion of an airport assessment which will be tailored to address specific shortfalls observed during the assessment, which could be effectively mitigated through training. These actions, when fully implemented, should address the intent of the recommendation. DHS concurred with the second recommendation that TSA establish criteria and guidance to assist TSA decision makers when determining the vulnerability rating of individual foreign airports. DHS stated that the most recent version of the *Foreign Airport Assessment Program Standard Operating Procedures* now contains several scenarios for managers to use as a set of guidelines in determining the vulnerability rating for each open standard and for the airport overall. DHS also stated that the Director of Global Compliance and ROC managers will collaborate on the development of a scenario archive to promote more long-term consistency in the event that key staff leave the agency. We support TSA's efforts to ensure greater transparency and consistency over how airport vulnerability scores are determined and believe it will be important for TSA to provide sufficient detail in the criteria and guidance that the agency develops. Such actions, when fully implemented, should address the intent of the recommendation. DHS concurred with the third recommendation that TSA consider the feasibility of conducting more targeted assessments and systematically compiling information on aviation security best practices. In its response, DHS stated that TSA is developing a pre-audit questionnaire that will be sent to each host government in advance of a planned airport assessment which will assist assessment teams in obtaining administrative information and key documents, such as the Airport Security Program, prior to the visit. DHS added that when the questionnaire is returned to TSA, the agency will obtain an official translation of all submitted items so that the assessment team has a better understanding of the current policies, procedures, and practices in place at the site. According to DHS, this practice may enable the team to tailor its efforts at the airports to focus on those areas of concern as indicated in the responses to the questionnaire, as well as the critical standards. DHS stated that TSA plans to complete development of the questionnaire by mid-fiscal year 2012, with wide-scale deployment beginning in October 2012. We support TSA's planned actions but also believe that there may be additional opportunities for TSA to expand its use of targeted assessments as it works to implement the first recommendation related to developing a mechanism to evaluate the results of completed assessment activities to determine any trends and target activities and resources. For example, as TSA works to systematically analyze the results of its assessments, it may determine that specific regions of the world need assistance in meeting certain critical standards. Such action, in conjunction with TSA's planned efforts, would meet the intent of the recommendation. With regard to aviation security best practices, DHS stated that the five volumes of the International Civil Aviation Organization (ICAO) *Security Manual for Safeguarding Civil Aviation Against Acts of Unlawful Interference* (Document 8973) contains the globally-recognized best practices and alternative methods for meeting the ICAO standards and recommended practices.

DHS stated that TSA participates in the development and review of this document and draws from it when recommending improvements to foreign airport authorities. However, it noted that an infrequently-populated portion of the foreign airport assessment reports is available for inspectors to capture particularly noteworthy practices.

DHS stated that during fiscal year 2012, inspectors will be encouraged to more conscientiously identify and document new approaches encountered at airports that are not reflected in the security manual but effectively address the ICAO standards and recommended practices. We support these efforts but also believe that it will be important for TSA to capture information identifying security best practices and technology that may be applicable to enhancing the security of U.S. airports. Such action, in conjunction with TSA's planned efforts, would meet the intent of the recommendation.

DHS also provided us with technical comments, which we incorporated as appropriate.

Stephen M. Lord
Director, Homeland Security and Justice Issues

APPENDIX I. SCOPE AND METHODOLOGY

To examine the efforts made by the Transportation Security Administration (TSA) to determine whether foreign airports that provide service to the United States are maintaining and carrying out effective security measures, we addressed the following questions: (1) to what extent has TSA taken steps to enhance its foreign airport assessment program since 2007, and what challenges remain; (2) what are the results of TSA's foreign airport assessments, and to what extent does TSA use the results of these assessments to guide its future assessment activities; and (3) what opportunities, if any, exist to enhance the value of TSA's foreign airport assessment program?

To collectively address all three questions, we reviewed relevant laws and regulations, including statutory provisions that identify specific actions to be taken by the Secretary of Homeland Security when the Secretary determines that a foreign airport does not maintain and carry out effective security measures.[53] We reviewed various TSA program management and strategic planning documents and interviewed TSA officials located at TSA headquarters and in the field. We interviewed other federal and nonfederal stakeholders, such as the Department of State, International Civil Aviation Organization (ICAO), and the European Commission (EC). We outline the specific steps taken to answer each objective below.

To determine the steps TSA has taken to enhance its foreign airport assessment program since 2007, we reviewed various TSA program management and strategic planning documents to identify revisions to its current and planned future strategy. Specifically, we reviewed TSA's 2010 *Foreign Airport Assessment Program Standard Operating Procedures* (SOP) document, which prescribes program and operational guidance for assessing security measures at foreign airports, and informs TSA personnel at all levels of what is expected of them in the implementation of the program. We also reviewed the job aids TSA inspectors use during each assessment, which ensure that the TSA-specified ICAO aviation security standards and recommended practices are fully evaluated during each assessment.

To determine TSA's current and planned future strategy, we reviewed available strategic planning documents that TSA uses to guide its program. Specifically, we reviewed TSA's Office of Global Strategies *International Strategy to Enhance Aviation Security* for 2010–2012, TSA's Office of Global Strategies Global Compliance Strategic Implementation Plan Fiscal Year 2011, and the TSA Capacity Development Strategic Plan for fiscal years 2011–2015. In addition, we also obtained and reviewed multilateral and bilateral arrangements TSA has established with the European Union (EU) and several foreign nations to facilitate coordination in the area of aviation security, including facilitation of TSA's foreign airport assessments.

To understand how TSA assesses and manages its foreign airport risk information, we obtained and reviewed various program documents. Specifically, we obtained and reviewed documents on TSA's methodology for assigning individual risk rankings (called Tier rankings) to each foreign airport it assesses. TSA's rankings are based on the likelihood of a location being targeted, the protective measures in place at that location, and the potential impact of an attack on the international transportation system. Airports are then categorized as high, medium, or low risk. While we did not evaluate the quality of TSA's risk rankings, as this analysis was outside the scope of our work, we generally determined that the rankings addressed all three components of risk (threat, vulnerability, and consequence).

To obtain a greater understanding of the foreign airport assessment process, including how TSA works with host nation officials, we accompanied a team of TSA inspectors during an assessment of the Toronto Pearson International Airport. We based our selection on several factors, including the airport locations TSA had plans to assess during the course of our audit work, host government willingness to allow us to accompany TSA, and travel costs.

To obtain information on the extent to which TSA provided oversight of its assessment efforts, we obtained and reviewed various TSA program management documents and tools TSA uses to track and manage information for the program. Specifically, we reviewed the TSA Airport and Air Carrier Comprehensive Tool (known as the A.C.T.), which TSA uses to track its foreign airport assessment schedule, including when various airports are due to be assessed.

We also reviewed the Open Standards and Recommended Practices Tracking Tool, which the TSA Representatives (TSAR) use to monitor and track a foreign airport's progress in resolving security deficiencies identified by TSA inspectors during previous assessments. In addition, we reviewed the tracking sheet TSA's Director of Global Compliance uses to compile and track current- and prior-year assessment results, including individual airport vulnerability scores and information on which specific ICAO standards were in noncompliance.

To obtain stakeholder views and perspectives on steps TSA has taken to enhance its foreign airport assessment program since 2007, we interviewed and obtained information from various federal and nonfederal stakeholders. Specifically, we interviewed TSA officials located in the Office of Global Strategies (OGS), Global Compliance (GC), Office of International Operations (OIO), and Capacity Development Branch (CDB).

In addition, we also conducted site visits to four of the five TSA Regional Operations Centers (ROC) located in Los Angeles, Dallas, Miami, and Frankfurt where we met with the ROC managers and 23 international aviation security inspectors who conduct TSA's foreign airport assessments.[54] We based our site visit selections on the number of available inspectors at each location and geographic dispersion.

We conducted telephone and in-person interviews with 9 of the 27 TSARs, located in various embassies and consulates throughout the world, who schedule TSA airport assessment visits and follow up on host governments' progress in addressing identified security deficiencies. When possible, we conducted in-person interviews with TSARs who were at TSA ROCs during our site visits.[55]

We based our TSAR selections on geographic dispersion and varying years of experience. During each of these interviews, we discussed these officials' responsibilities related to the program, including their role in assisting foreign officials in correcting security deficiencies identified during assessments.

We met with Department of State officials to better understand how they coordinate with TSA through their Anti-Terrorism Assistance (ATA) Program and other related efforts aimed at assisting foreign partners' capacity to secure their airports.

Additionally, we met with officials from the EC, International Air Transport Association, and ICAO to discuss efforts and programs these organizations have in place to enhance international aviation security. We interviewed or received responses to questions from five foreign embassies to obtain perspectives of foreign transportation security officials on TSA's airport assessment program.

We based our selection on geographic dispersion and countries with the highest risk airports, as designated by TSA.[56] However, information from our interviews with government officials, members of the aviation industry, and

TSA officials and inspectors cannot be generalized beyond those that we spoke with because we did not use statistical sampling techniques in selecting individuals to interview.

To identify challenges affecting TSA's foreign airport assessment program, we interviewed TSA program management officials and field officials located at the TSA ROCs on the challenges they experience obtaining access to foreign airports to conduct assessments, the development of an automated database management system, and the provision of aviation security training to foreign governments.

In addition, we met with TSA's Director of Global Compliance, and ROC managers and inspectors located in the field, to discuss potential future challenges TSA may experience when attempting to conduct assessments at foreign airports with all-cargo flights to the United States. Specifically, we obtained their perspectives on foreign governments that have been reluctant to allow TSA inspectors to visit their airports.

We interviewed TSA's Director of Global Compliance on the agency's progress in developing an automated database to manage program information, including the challenges the agency has experienced finding a solution that meets program needs. We conducted telephone and in-person interviews with nine TSARs to obtain their perspectives on challenges to scheduling airport assessment visits.

In addition, we interviewed officials within TSA's CDB to better understand the scope and types of requests for assistance they receive from foreign countries, including challenges they experience in attempting to provide assistance, such as resource constraints and aligning security priorities with the Department of State.

To determine the results of TSA's foreign airport assessments and the extent to which the agency evaluates its results to inform future activities, we interviewed TSA officials on the results of its assessments, obtained and reviewed assessment reports and relevant program documents, and conducted our own independent analysis of TSA's assessment results.

To better understand the scope and type of information contained in TSA's foreign airport assessment reports, we obtained and reviewed the most recently available assessments for all high-risk airports. We also selected a randomized sample of assessment reports from current and prior years. We reviewed sections of these reports for completeness and general consistency with TSA guidance for preparing assessment reports.

We obtained and reviewed TSA's foreign airport risk-ranking sheet to better understand which airports TSA identified as high, medium, and low risk, including how the results of TSA's assessments influence an airport's risk ranking.

In addition, we obtained and reviewed TSA's foreign airport assessment program vulnerability results tracking sheet used by the Director of Global Compliance to compile and track current and prior- year assessment results. This tracking sheet included records of TSA's compliance assessments for each airport that TSA assessed from fiscal year 1997 through May 9, 2011. Specifically, the tracking sheet recorded assessment results for each of the ICAO standards used in the airport assessments, as well as an overall vulnerability score of 1 through 5 assigned after each assessment.

This overall vulnerability score is a representation of compliance or noncompliance with all the ICAO standards against which TSA assesses foreign airports. We interviewed the Director of Global Compliance on the steps taken to develop the tracking sheet, including how TSA manages and updates data, and how TSA assigns vulnerability scores.

In addition, we conducted our own independent analysis of TSA's assessment results from fiscal year 2006 through May 9, 2011. Specifically, we analyzed data from TSA's foreign airport assessment program vulnerability results tracking sheet to identify the number of airports in each vulnerability category by region.

We also analyzed TSA assessment results data to determine the frequency with which foreign airports complied with particular ICAO standards, such as access control, quality control, passenger screening, and baggage screening, among others. For those airports that TSA has identified as high risk, we analyzed TSA assessment results data to determine the number of resolved and remaining compliance issues at high-risk airports by region, as well as the level of noncompliance found at high-risk airports.

To assess the reliability of TSA's data, we selected a random sample of records from TSA's foreign airport assessment program vulnerability results tracking sheet. Next, we examined the corresponding reports to locate those ICAO standards that had been identified as less than fully compliant in the tracking sheet (a score of 2 through 5 on a 5-point scale).[57]

The actual scores assigned to the compliance ratings and found in the tracking sheet were determined by the Director of Global Compliance using guidance in the 2010 SOP in consultation with individuals involved in the assessment process (ROC managers, Supervisory Transportation Security Specialists, and Transportation Security Specialists).

Our comparison of the results in the tracking sheet with the compliance information provided in the corresponding reports did not match in several cases.

However, in discussions with TSA we determined that the differences were the result of changes to the ICAO standards used in the assessments or a change in the definition of the standards. Specifically, TSA told us that Amendments 10 and 11 to ICAO Annex 17 changed the definitions of some standards, and the numbers assigned to identify them. For example, a standard concerning Hold Baggage Security is now identified as 4.5.1. However, in years prior to Amendments 10 and 11 to Annex 17, that same standard was identified as 4.1.1.

TSA's Director of Global Compliance told us that she updated the foreign airport assessment program vulnerability results tracking sheet with the new definitions and numbers, and the associated results, each time an ICAO amendment came out. As a result, we determined that any analysis of the assessment results for specific ICAO standards would need to take into account the changes TSA identified.

Based on our overall analysis of the data and reports, we determined that the data were sufficiently reliable to provide a general indication, by type or category, of the standards TSA assesses against and the level of compliance, and frequency of compliance, for TSA's airport assessments over the period of our analysis.

In addition, we interviewed TSA's Director of Global Compliance on the steps TSA takes to analyze its assessment results to inform the agency's future efforts and compared these efforts to *Standards for Internal Control in the Federal Government*.[58]

We discussed the status of implementation of our 2007 recommendation to develop outcome-oriented performance measures to evaluate the impact that TSA assessments have on improving foreign airport compliance with ICAO standards. We interviewed TSA managers and inspectors located in the field on their roles and responsibilities in determining and documenting assessment results. We assessed TSA's efforts to analyze its assessment results against *Standards for Internal Control in the Federal Government*, which require agencies to ensure that ongoing monitoring occurs during the course of normal operations to help evaluate program effectiveness.

To identify opportunities for TSA to enhance the value of TSA's foreign airport assessment program, we reviewed all relevant program management and strategic documentation, and interviewed TSA officials as well as various other federal and nonfederal stakeholders. Specifically, we reviewed the 2011

Foreign Airport Assessment Program SOP and job aids; OGS, GC, and CDB strategic planning documents; foreign airport risk assessment and ranking information; program management tools TSA uses to track and manage its schedule and the status of foreign airport security deficiencies; and reviewed TSA foreign airport assessment results and reports. We also reviewed our prior work concerning how risk-informed and priority driven decisions can help inform agency decision makers in allocating finite resources to the areas of greatest need.[59]

Moreover, we reviewed the process TSA uses to assign vulnerability ratings of 1-5 to each foreign airport it assesses and then evaluated this process against *Standards for Internal Control in the Federal Government*, which call for controls and other significant events to be clearly documented in directives, policies, or manuals to help ensure operations are carried out as intended.[60] In addition, we visited the Toronto Pearson International Airport to observe TSA inspectors during the assessment thereby obtaining a greater understanding of the foreign airport assessment process, including opportunities for TSA to improve its program.

We reviewed prior GAO work discussing the importance of identifying potential best practices, as part of conducting U.S. federal government security assessments in other countries.

To obtain stakeholder views and perspectives on opportunities to enhance the program, we interviewed and obtained information from various TSA and nonfederal stakeholders. Specifically, we interviewed TSA headquarters officials in GC, OIO, and CDB. During our site visits, we interviewed ROC managers and international inspectors on possible opportunities that exist for TSA to improve its foreign airport assessment program. We discussed opportunities to improve the program during our telephone and in-person interviews with nine TSARs. In addition, we discussed ways in which TSA could improve its program during our interviews with officials from the EC, ICAO, and select foreign embassies.

We conducted this performance audit from August 2010 through October 2011 in accordance with generally accepted government auditing standards. Those standards require that we plan and perform the audit to obtain sufficient, appropriate evidence to provide a reasonable basis for our findings and conclusions based on our audit objectives.

We believe that the evidence obtained provides a reasonable basis for our findings and conclusions based on our audit objectives.

APPENDIX II. PROCESS FOR TAKING SECRETARIAL ACTION AGAINST A FOREIGN AIRPORT

If the Secretary, based on the TSA airport assessment results, determines that a foreign airport does not maintain and carry out effective security measures, he or she must, after advising the Secretary of State, take secretarial action. Below is a list of these actions.

Figure 4 describes the process for taking secretarial action against an airport.

- **90-day action**—The Secretary notifies foreign government officials that they have 90 days to address security deficiencies that were identified during the airport assessment and recommends steps necessary to bring the security measures at the airport up to ICAO standards.[61]
- **Public notification**—If, after 90 days, the Secretary finds that the government has not brought security measures at the airport up to ICAO standards, the Secretary notifies the general public that the airport does not maintain and carry out effective security measures.[62]
- **Modification to air carrier operations**—If, after 90 days, the Secretary finds that the government has not brought security measures at the airport up to ICAO standards:
  - The Secretary may withhold, revoke, or prescribe conditions on the operating authority of U.S.-based and foreign air carriers using that airport to provide transportation to the U.S., following consultation with appropriate host government officials and air carrier representatives, and with the approval of the Secretary of State.[63]
  - The President may prohibit a U.S.-based or foreign air carrier from providing transportation between the United States and any foreign airport that is the subject of a secretarial determination.[64]
- Suspension of service— The Secretary, with approval of the Secretary of State, shall suspend the right of any U.S.-based or foreign air carrier to provide service to or from an airport if the Secretary determines that a condition exists that threatens the safety or security of passengers, aircraft, or crew traveling to or from the airport, and the public interest requires an immediate suspension of transportation between the United States and that airport.[65]

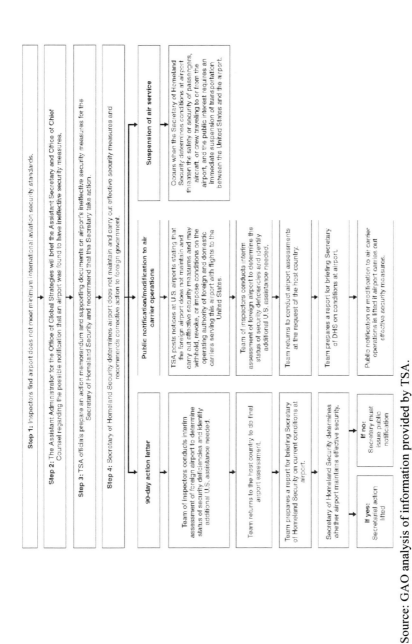

Source: GAO analysis of information provided by TSA.

Figure 1. Process for Taking Secretarial Action against a Foreign Airport.

APPENDIX III. ICAO STANDARDS TSA USES TO ASSESS SECURITY MEASURES AT FOREIGN AIRPORTS

TSA inspectors use 40 ICAO standards and 1 recommended practice when conducting foreign airport assessments. Of the 40, TSA identified 22 as critical. These 22 critical standards are in bold.[66]

Airport Operations

3.2.1. Each Contracting State shall require each airport serving civil aviation to establish, implement and maintain a written Airport Security Program appropriate to meet the requirements of the National Civil Aviation Security Programme.

3.2.2. Each Contracting State shall ensure that an authority at each airport serving civil aviation is responsible for coordinating the implementation of security controls.

3.2.3. Each Contracting State shall ensure that an airport security committee at each airport serving civil aviation is established to assist the authority mentioned under 3.2.2 in its role of coordinating the implementation of security controls and procedures as specified in the airport security programme.

Quality Control

3.4.1. Each Contracting State shall ensure that the persons implementing security controls are subject to background checks and selection procedures.

3.4.2. Each Contracting State shall ensure that the persons implementing security controls possess all competencies required to perform their duties and are appropriately trained according to the requirements of the national civil aviation security programme and that appropriate records are maintained up to date. Relevant standards of performance shall be established and initial and periodic assessments shall be introduced to maintain those standards.

3.4.3. Each Contracting State shall ensure that the persons carrying out screening operations are certified according to the requirements of the

National Civil Aviation Security Program to ensure that performance standards are consistently and reliably achieved.

3.4.5. Each Contracting State shall ensure that the implementation of security measures is regularly subjected to verification of compliance with the national civil aviation security programme. The priorities and frequency of monitoring shall be determined on the basis of risk assessment carried out by the relevant authorities.

3.4.6. Each Contracting State shall arrange for audits, tests, surveys and inspections to be conducted on a regular basis, to verify compliance with the National Civil Aviation Security Program and to provide for the rapid and effective rectification of any deficiencies.

Measures Relating to Access Control

4.2.1. Each Contracting State shall ensure that the access to airside areas at airports serving civil aviation is controlled in order to prevent unauthorized entry.

4.2.2. Each Contracting State shall ensure that security restricted areas are established at each airport serving civil aviation designated by the State based upon a security risk assessment carried out by the relevant national authorities.

4.2.3. Each Contracting State shall ensure that identification systems are established in respect of persons and vehicles in order to prevent unauthorized access to airside areas and security restricted areas. Identity shall be verified at designated checkpoints before access is allowed to airside areas and security restricted areas.

4.2.4. Each Contracting State shall ensure that background checks are conducted on persons other than passengers granted unescorted access to security restricted areas of the airport prior to granting access to security restricted areas.

4.2.5. Each Contracting State shall ensure that the movement of persons and vehicles to and from the aircraft is supervised in security restricted areas in order to prevent unauthorized access to aircraft.

4.2.6. Each Contracting State shall ensure that persons other than passengers, together with items carried, being granted access to security restricted areas are screened; however, if the principle of 100 per cent screening cannot be accomplished, other security controls, including but not limited to proportional screening, randomness and unpredictability, shall be applied in accordance with a risk assessment carried out by the relevant national authorities.

4.2.7. Each Contracting State shall ensure that vehicles being granted access to security restricted areas, together with items contained within them, are subject to screening or other appropriate security controls in accordance with a risk assessment carried out by the relevant national authorities.

Measures Relating to Aircraft

4.3.1. Each Contracting State shall ensure that aircraft security checks of originating aircraft engaged in commercial air transport movements are performed or an aircraft security search is carried out. The determination of whether it is an aircraft security check or a search that is appropriate shall be based upon a security risk assessment carried out by the relevant national authorities.

4.3.2. Each Contracting State shall ensure that measures are taken to ensure that any items left behind by passengers disembarking from transit flights are removed from the aircraft or otherwise dealt with appropriately before departure of an aircraft engaged in commercial flights.

4.3.3. Each Contracting State shall require its commercial air transport operators to take measures as appropriate to ensure that during flight unauthorized persons are prevented from entering the flight crew compartment.

4.3.4. Each Contracting State shall ensure that an aircraft subject to 4.3.1 is protected from unauthorized interference from the time the aircraft search or check has commenced until the aircraft departs.

Measures Relating to Passengers and Their Cabin Baggage

4.4.1. Each Contracting State shall establish measures to ensure that originating passengers of commercial air transport operations and their cabin baggage are screened prior to boarding an aircraft departing from a security restricted area.

4.4.2. Each Contracting State shall ensure that transfer passengers of commercial flights and their cabin baggage are screened prior to boarding an aircraft, unless it has established a validation process and continuously implements procedures, in collaboration with the other Contracting State where appropriate, to ensure that such passengers and their cabin baggage have been screened to an appropriate level at the point of origin and, subsequently, protected from unauthorized interference from the point of screening at the originating airport to the departing aircraft at the transfer airport.

4.4.3. Each Contracting State shall ensure that passengers and their cabin baggage which have been screened are protected from unauthorized interference from the point of screening until they board their aircraft. If mixing or contact does take place, the passengers concerned and their cabin baggage shall be re-screened before boarding an aircraft.

4.4.4. Each Contracting State shall ensure that passengers and their cabin baggage which have been screened are protected from unauthorized interference from the point of screening until they board their aircraft. If mixing or contact does take place, the passengers concerned and their cabin baggage shall be re-screened before boarding an aircraft.

Measures Relating to Hold Baggage

4.5.1. Each Contracting State shall establish measures to ensure that originating hold baggage is screened prior to being loaded onto an aircraft engaged in commercial air transport operations departing from a security restricted area.

4.5.2. Each Contracting State shall ensure that all hold baggage to be carried on a commercial aircraft is protected from unauthorized interference from the point it is screened or accepted into the care of the carrier, whichever is earlier, until departure of the aircraft on which it is to be carried. If the integrity of the hold baggage is jeopardized, the hold baggage shall be re-screened before being placed on board an aircraft.

4.5.3. Each Contracting State shall ensure that commercial air transport operators do not transport the baggage of passengers who are not on board the aircraft unless that baggage is identified as unaccompanied and subjected to additional screening.

4.5.4. Each Contracting State shall ensure that transfer hold baggage is screened prior to being loaded onto an aircraft engaged in commercial air transport operations, unless it has established a validation process and continuously implements procedures, in collaboration with the other Contracting State where appropriate, to ensure that such hold baggage has been screened at the point of origin and subsequently protected from unauthorized interference from the originating airport to the departing aircraft at the transfer airport.

4.5.5. Each Contracting State shall ensure that aircraft commercial air transport operators transport only items of hold baggage that have been individually identified as accompanied or unaccompanied, screened to the appropriate standard, and accepted for carriage on that flight by the air carrier. All such baggage should be recorded as meeting these criteria and authorized for carriage on that flight.

Measures Relating to Cargo, Mail and Other Goods

4.6.1. Each Contracting State shall ensure that appropriate security controls, including screening where practicable, are applied to cargo and mail, prior to their being loaded onto an aircraft engaged in passenger commercial air transport operations.

4.6.2. Each Contracting State shall establish a supply chain security process, which includes the approval of regulated agents and/or known consignors, if such entities are involved in implementing screening or other security controls of cargo and mail.

4.6.3. Each Contracting State shall ensure that cargo and mail to be carried on a passenger commercial aircraft are protected from unauthorized interference from the point screening or other security controls are applied until departure of the aircraft.

4.6.4. Each Contracting State shall ensure that operators do not accept cargo or mail for carriage on an aircraft engaged in passenger commercial air transport operations unless the application of screening or other security controls is confirmed and accounted for by a regulated agent, or such consignments are subjected to screening. Consignments which cannot be confirmed and accounted for by a regulated agent are to be subjected to screening.

4.6.5. Each Contracting State shall ensure that catering, stores and supplies intended for carriage on passenger commercial flights are subjected to appropriate security controls and thereafter protected until loaded onto the aircraft.

4.6.6. Each Contracting State shall ensure that merchandise and supplies introduced into security restricted areas are subject to appropriate security controls, which may include screening.

4.6.7. Each Contracting State shall ensure that security controls to be applied to cargo and mail for transportation on all-cargo aircraft are determined on the basis of a security risk assessment carried out by the relevant national authorities.

Measures Relating to Special Categories of Passengers

4.7.1. Each Contracting State shall develop requirements for air carriers for the carriage of potentially disruptive passengers who are obliged to travel because they have been the subject of judicial or administrative proceedings.

Measures Relating to the Landside

4.8.1. Recommendation.— Each Contracting State should ensure that security measures in landside areas are established to mitigate possible threats of acts of unlawful interference in accordance with a risk assessment carried out by the relevant authorities.

Prevention

5.1.4. Each Contracting State shall ensure that contingency plans are developed and resources made available to safeguard civil aviation, against acts of unlawful interference. The contingency plans shall be tested on a regular basis.

5.1.5. Each Contracting State shall ensure that authorized and suitably trained personnel are readily available for deployment at its airports serving international civil aviation to assist in dealing with suspected, or actual, cases of unlawful interference with civil aviation.

Aerodrome Emergency Planning

9.1.1. An aerodrome emergency plan shall be established at an aerodrome, commensurate with the airport operations and other activities conducted at the aerodrome.

Fencing

9.10.3. Suitable means of protection shall be provided to deter the inadvertent or premeditated access of unauthorized persons into ground installations and facilities essential for the safety of civil aviation located off the aerodrome.

APPENDIX IV. TSA PROCESS FOR CONDUCTING FOREIGN AIRPORT ASSESSMENTS

TSA uses a multistep process to conduct its assessments of foreign airports. Figure 5 describes the process TSA uses.

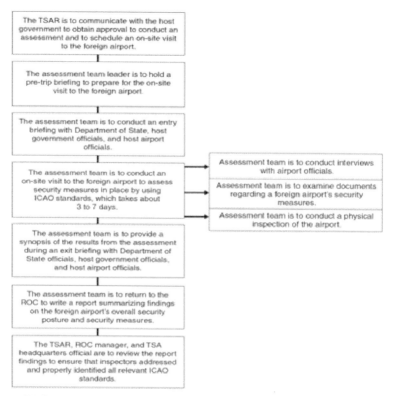

Source: GAO analysis of information provided by TSA.

Figure 2. Multistep Process for Conducting Foreign Airport Assessments.

APPENDIX V. TSA AVIATION SECURITY SUSTAINABLE INTERNATIONAL STANDARDS TEAM (ASSIST) PROGRAM

The mission of the ASSIST program is to raise and strengthen international aviation security standards in foreign countries and airports, and to ensure that improvements in standards are long-term and sustainable. Specifically, TSA deploys teams consisting of six to seven individuals for 1 week in partnership with the host nation in order to evaluate and develop recommendations for building the aviation security capacity. Following the initial visit, TSA conducts follow-up focused visits to deliver training and technical assistance when agreed upon by the host nation. To date, TSA has partnered with five foreign countries under the ASSIST program.[67] These countries are St. Lucia, Liberia, Georgia, Haiti, and Palau. TSA selects countries to partner with based on a variety of factors, which include focusing on countries with last point of departure service to the United States, foreign airport risk rankings, a foreign government's demonstrated willingness to engage TSA, and a foreign government's demonstrated ability to sustain ASSIST initiatives after the conclusion of ASSIST. See below for specific information on the countries TSA partnered with during 2009-2011.

St. Lucia

St. Lucia was the first nation to partner with TSA under the ASSIST program. It was selected as the pilot country for ASSIST because it is a last point of departure location to the U.S., a popular destination for U.S. passengers, and the TSA Representative in the region requested the assistance. The inaugural survey visit to St. Lucia was conducted in January 2009. Subsequent follow-up visits were held in March and June of 2009, and focused on training in Emergency Communications, Improvised Explosive Device Familiarization, Essential Instructor Skills, and Basic Screener Training. The ASSIST program closed out in St. Lucia in 2010. TSA officials told us that TSA partnered with St. Lucia because it was the pilot country for the ASSIST program. The Capacity Development Branch did not want to pilot the ASSIST program in a country that was "ultra challenging" in terms of security deficiencies.

Liberia

Liberia was the second nation to partner with TSA under ASSIST. Liberia was chosen for ASSIST after President George W. Bush visited the nation in February 2008, pledging U.S. support in the area of aviation security. In addition, Delta Airlines wanted to reestablish service between the U.S. and Liberia and, in order to do so, Liberia's national civil aviation program needed improvement. Liberia received a survey visit in April 2009. TSA conducted Essential Instructor Skills and Basic Screening Skills Training in May 2009. This training was followed by monthly visits to assess the impact of training and other technical assistance. In January 2010, TSA coordinated Fraudulent Document Detection training in conjunction with the U.S. Customs and Border Protection and Immigration and Customs Enforcement. In August 2010, TSA conducted its National Inspectors Training. The ASSIST program was closed out in Liberia in November 2010.

Georgia

Georgia, the third nation to partner with TSA under ASSIST, received a survey visit in September 2009. TSA coordinated its ASSIST program activities in Georgia with the European Civil Aviation Conference (ECAC). Georgia is a member state of ECAC and ECAC initiated a program of technical assistance in Georgia following its March 2009 audit of the Tbilisi Airport. In addition, TSA officials also told us that the State Department also requested that TSA work with Georgia. In April 2010, ECAC and TSA conducted ECAC's Best Practices for National-level Auditors course. In August 2010, TSA conducted a review of passenger and baggage screening. The ASSIST program was closed out in Georgia in December 2010.

Palau

TSA deployed an ASSIST program representative to Palau in August 2010. TSA officials told us that Palau was selected for the ASSIST program as a result of the results from the TSA foreign airport assessment program. In addition, Palau was a last point of departure to the United States and the host government was willing to engage TSA and make a commitment to sustain its aviation security enhancements.

Haiti

Currently, the ASSIST program is working with Haiti. Haiti was selected for ASSIST as a result of past program assessment recommendations. Specifically, in October 2010, the ASSIST team was in the process of conducting a training "needs assessment" in Haiti to determine what is needed to rectify aviation security deficiencies found by the program.

End Notes

[1] See 49 U.S.C. § 114(d).

[2] See 49 U.S.C. § 44907.

[3] GAO, *Aviation Security: Foreign Airport Assessments and Air Carrier Inspections Help Enhance Security, but Oversight of These Efforts Can Be Strengthened*, GAO-07-729 (Washington, D.C.: May 11, 2007).

[4] GAO, *Standards for Internal Control in the Federal Government*, GAO/AIMD-00-21.3.1 (Washington, D.C.: November 1999).

[5] See Pub. L. No. 107-71, 115 Stat. 597 (2001).

[6] 49 U.S.C. § 44907. Prior to the establishment of DHS in March 2003, authority for conducting foreign airport assessments resided with the Secretary of Transportation. Although assessments were originally conducted by the Federal Aviation Administration (FAA), TSA assumed responsibility for conducting the assessments following the enactment of the Aviation and Transportation Security Act (enacted Nov. 19, 2001). In March 2003, TSA transferred from the Department of Transportation to DHS. See Homeland Security Act of 2002, Pub. L. No. 107-296, § 403(2), 116 Stat. 2135, 2178.

[7] See 49 U.S.C. § 44907(d)-(e).

[8] Domestic and foreign air carriers that operate to, from, or within the United States must establish and maintain security programs approved by TSA in accordance with requirements set forth in regulation at 49 C.F.R. parts 1544 and 1546. See 49 U.S.C §§ 44903, 44906. As with foreign airport assessments, FAA had responsibility for conducting air carrier inspections prior to TSA's establishment and assumption of this function.

[9] ICAO was formed following the 1944 Convention on International Civil Aviation (also known as the Chicago Convention). In 1947, ICAO became a specialized agency of the United Nations. A primary objective of ICAO is to provide for the safe, orderly, and efficient development of international civil aviation. There are currently 190 signatory nations to the ICAO convention, including the United States. Nations that are members to the ICAO convention agree to cooperate with other member states to meet standardized international aviation security measures. The international aviation security standards and recommended practices are detailed in Annex 17 to the Convention on International Civil Aviation adopted by ICAO.

[10] More specifically, an ICAO standard is a specification for the safety or regularity of international air navigation, with which member states agree to comply; a recommended practice is any desirable specification for safety, regularity, or efficiency of international air navigation, with which member states are strongly encouraged to comply. Member states are expected to make a genuine effort to comply with recommended practices.

[11] Segments of Annex 17 to the Convention of International Civil Aviation, *Safeguarding International Civil Aviation Against Unlawful Acts of Interference*, Ninth Edition, March

2011, and Annex 14, *Aerodrome Design and Operations*, Volume I, have been reproduced in appendix III with permission of the International Civil Aviation Organization.

[12] A risk-informed approach entails consideration of terrorist threats, vulnerability of potential terrorist targets to those threats, and the consequences of those threats being carried out when deciding how to allocate resources to defend against these threats. Risk- informed decision making can help ensure that finite resources are allocated to the areas of greatest need.

[13] According to TSA, the airport assessment period is extended by 8 to 12 hours for each air carrier inspection that TSA conducts in conjunction with an airport assessment. TSA may also conduct air carrier inspections separately from airport assessments because foreign airports are generally assessed no more than once a year by TSA, while some air carriers are inspected twice a year by TSA. See 49 C.F.R. §§ 1544.3, 1546.3 (requiring that each U.S. aircraft operator and foreign air carrier (respectively) allow TSA, at any time or place, to make any inspections or tests to determine compliance-applicable requirements).

[14] TSA requires that each air carrier adopt and implement a TSA-approved security program for all scheduled passenger and public charter operations at locations within the United States, from the United States to a non-U.S. location, or from a non-U.S. location to the United States. See 49 C.F.R. pts. 1544-46. When TSA determines that additional security measures are necessary to respond to a threat assessment or to a specific threat against civil aviation, TSA may issue a security directive or an emergency amendment to an air carrier security program that sets forth additional mandatory security requirements. Air carriers are required to comply with each applicable security directive or emergency amendment issued by TSA, along with the requirements already within their security programs and any other requirements set forth in applicable law.

[15] According to TSA's *Foreign Airport Assessment Program Standard Operating Procedures*, if security concerns and deficiencies are considered not "serious enough to warrant secretarial action (e.g., the measure barely satisfies the minimum international standard and could be improved)," TSA may develop an action plan for addressing the deficiencies identified without seeking a determination from the Secretary of Homeland Security.

[16] TSA officials told us they used their subject-matter expertise and expert judgment to identify the 40 standards, which allow them to focus only on areas most critical for their assessments. TSA also assesses foreign airports against one ICAO-recommended practice concerning landslide areas. See appendix III for the complete list of standards TSA assesses foreign airports against.

[17] These 22 standards cover the areas of passenger and hold baggage screening, access control, aircraft-in-flight security, and cargo/catering/mail.

[18] The Director of Global Compliance told us that if multiple trips are scheduled back-to- back, inspectors are to conduct the air carrier inspection visit at one airport first, and the airport visit combining the assessment and air carrier inspection second. Doing so provides the inspectors more time to meet the 20 day airport assessment reporting requirement.

[19] Of the 25 new international inspector positions, 5 are allocated for Frankfurt ROC; 6 for Miami; 7 for Dallas; 2 for Singapore; and 5 for Los Angeles.

[20] TSA is also planning to hire a program analyst in headquarters to assist with, among other things, analyzing assessment results and assessments scheduling. TSA also established three new TSAR positions in 2010—one each in Brasilia, Brazil; Johannesburg, South Africa; and Nassau, the Bahamas; and plans to staff three more TSARs by the end of 2011 located in Dakar, Senegal; Dubai, United Arab Emirates; and New Delhi, India.

[21] A risk-informed approach entails consideration of terrorist threats, vulnerability of potential terrorist targets to those threats, and the consequences of those threats being carried out when deciding how to allocate resources to defend against these threats. Risk- informed, priority-driven decisions can help inform decision makers in allocating finite resources to the areas of greatest need.

[22] The total number of foreign airports TSA assesses changes due to carriers changing service locations or flight destinations, and seasonal service carriers.

[23] Information on the specific number of airports identified as high, medium and low risk is deemed sensitive security information.

[24] Under the previous approach, foreign airports that exhibited no operational issues in the previous two assessments were assessed once every 3 years. Foreign airports that had not been previously assessed, were subjected to secretarial action within the last 5 years, or exhibited operational issues in either of the two previous assessments were assessed once a year. Operational issues were considered weaknesses in the security system at an airport that pose a direct threat to the safety and security of passengers, aircraft, and crew (i.e., screening and access control measures).

[25] GAO, *Transportation Security: Comprehensive Risk Assessments and Stronger Internal Controls Needed to Help Inform TSA Resource Allocation*, GAO-09-492 (Washington, D.C.: Mar. 27, 2009).

[26] TSA also plans to develop additional 1 year implementation plans for future years, which will also include long-term foreign airport assessment program goals and objectives.

[27] These courses consist of Preventive Security Measures, Incident Management and Response, Excellence in Screening Techniques, Cargo Security Management, Essential Instructor Skills, and a National Inspectors Workshop. Additional courses in the process of development include Instructional Systems Design, and National Programs: National Civil Aviation Security Program, National Civil Aviation Security Quality Control Program, National Civil Aviation Security Training Program, Information Gathering through Casual Conversation, and Incident Management and Response.

[28] In the homeland security context, "harmonization" is a broad term used to describe countries' efforts to coordinate their security practices to enhance security and increase efficiency by avoiding duplication of effort. Harmonization efforts can include countries' mutually recognizing and accepting each other's existing practices—which could represent somewhat different approaches to achieve the same outcome, as well as working to develop uniform standards.

[29] The European Commission is the executive body of the European Union. The body is responsible for proposing legislation, implementing decisions, upholding the Union's treaties and the general day-to-day running of the Union. The Commission operates as a cabinet government, with 27 commissioners (one commissioner per member state). The Commission is required to monitor Member States' compliance with the aviation security legislation and carries out inspections of national appropriate authorities, airport inspections, and follow-up inspections to confirm the implementation of remedial actions.

[30] The 27 member states of the European Union are Austria, Belgium, Bulgaria, Czech Republic, Denmark, Germany, Estonia, Greece, Spain, France, Ireland, Italy, Cyprus, Latvia, Lithuania, Luxembourg, Hungary, Malta, the Netherlands, Poland, Portugal, Romania, Slovenia, Slovakia, Finland, Sweden, and the United Kingdom.

[31] According to TSA officials, if a table-top review does not provide TSA with sufficient information to make a determination on the security posture of the airport, TSA will conduct an independent assessment.

[32] See 73 Fed. Reg. 53,034 (Sept. 12, 2008). The notice directed all U.S. and foreign air carriers (and their agents) providing service between the United States and Venezuelan airports, to provide notice to any passenger purchasing a ticket for transportation between the United States and these airports that DHS is unable to determine whether such airports maintain and carry out effective security measures, and further required that similar notices be posted at U.S. airports. The notice remains in effect.

[33] TSA security requirements for U.S.-bound flights cover critical areas of aviation security including passenger, baggage, and cargo screening.

[34] AIT produces an image of a passenger's body that a screener interprets. The image identifies metallic and nonmetallic threats including weapons, explosives, and other objects concealed under layers of clothing.

[35] The State Department's ATA program seeks to provide partner countries the training, equipment, and technology they need to combat terrorism and prosecute terrorists and terrorist supporters. The Anti-Terrorism Assistance program was established in 1983.

[36] The Organization of American States is made up of 35 member states, including the independent nations of North, Central, and South America and the Caribbean, and is a forum for strengthening democracy, promoting human rights, and confronting shared problems among its members, such as poverty, terrorism, illegal drugs, and corruption.

[37] Its eight employees comprise the CDB Manager that oversees its International Instructions Development and Design group—two instructional systems design managers; an International Instructional Delivery group—three program analysts/instructors; and an International Technical Assistance group—two program managers for its ASSIST program.

[38] Information on the specific number of countries with high-risk airports as identified by TSA is deemed SSI.

[39] Specifically, by the end of fiscal year 2011, TSA is to finalize a coordinated set of procedures for the lending of decommissioned expendable and nonexpendable TSA aviation security equipment to partner countries, and develop a risk-based methodology for lending such equipment. In addition, by the end of fiscal year 2013, TSA is to implement the risk-based methodology through TSARs to the prioritized list of equipment recipients, and implement an evaluation plan to determine program impact.

[40] See 49 U.S.C. § 114(m) (referencing § 106(l), (m)).

[41] TSA also stated that it may also provide equipment deemed excess or surplus to foreign governments in accordance with General Services Administration guidance and regulations.

[42] The dates of deployment were October 31-November 12, 2010, November 28- December 3, 2010, and January 21-25, 2011.

[43] This includes foreign all-cargo operation airports as well as all-cargo flights departing from airports that also provide passenger service to countries other than the United States.

[44] According to TSA, the number of airports to which the agency assigns a risk ranking and that are therefore eligible for assessment, is constantly in flux, as air carriers start and stop service to the U.S. from foreign locations for a variety of reasons, such as seasonal service. While, as of January 1, 2011, TSA has categorized 277 foreign airports, TSA officials told us that the number of airports eligible for assessment typically ranges from about 275 to 300. In addition, our analyses are based primarily on data provided by TSA on May 9, 2011. At that time, there were 35 ICAO standards against which TSA assessed airports, including 17 critical standards. The data also included assessments on 1 additional standard that had been used in previous fiscal years but was no longer active on May 9, 2011.

[45] For example, this tool provides each airport's vulnerability category, or score, and includes information on the frequency with which each foreign airport complied with particular categories of ICAO standards, such as passenger screening, checked baggage screening, and access controls, among others.

[46] GAO.

[47] GAO.

[48] Pub. L. No. 111-352, 124 Stat. 3866 (2011) (amending Pub. L. No. 103-62, 107 Stat. 285 (1993)). See also 31 U.S.C. § 1115 (relating to performance measurement).

[49] "Leading security indicators" refers to an airport's vulnerability rating for its security posture, ranked on a scale of 1 (fully compliant) to 5 (egregious noncompliance).

[50] GAO.

[51] GAO, *Aviation Security: Federal Efforts to Secure U.S.-Bound Air Cargo Are in the Early Stages and Could Be Strengthened*, GAO-07-660 (Washington, D.C.: Apr. 30 2007).

[52] GAO.

[53] See, 49 U.S.C. § 44907.

[54] We did not visit the Singapore ROC due to travel costs and the small number of TSA inspectors at this location.

[55] We conducted in-person interviews with TSARs during our site visits to Miami and Frankfurt.

[56] These embassies included Canada, Mexico, France, United Kingdom, and Australia.

[57] Since the actual numerical scores were not recorded in the assessment reports, it was only possible to identify those standards that were identified as not fully compliant (i.e., standards which corresponded to a score greater than 1 on the tracking sheet).

[58] GAO, *Standards for Internal Control in the Federal Government*, GAO/AIMD-00-21.3.1 (Washington, D.C.: November 1999).

[59] GAO, *Aviation Security: Federal Efforts to Secure U.S.-Bound Air Cargo Are in the Early Stages and Could Be Strengthened*, GAO-07-660 (Washington, D.C.: Apr. 30 2007).

[60] GAO

[61] The Secretary may bypass the 90-day action and immediately provide public notification or withhold, revoke, or prescribe conditions on an air carrier's operating authority if the Secretary determines, after consultation with the Secretary of State, that a condition exists that threatens the safety or security of passengers, aircraft, or crew traveling to or from the airport. § 44907(d)(2)(A)(ii).

[62] Public notification includes publication of the airport's identity in the Federal Register, posting and displaying the airport's identity prominently at all U.S. airports at which scheduled air carrier operations are provided regularly, and notifying news media of the airport's identity. 49 U.S.C. § 44907(d)(1)(A). U.S. and foreign air carriers providing transportation between the United States and the airport shall also provide written notice that the airport is not maintaining and carrying out effective security measures on or with the ticket to each passenger buying a ticket. § 44907(d)(1)(B).

[63] § 44907(d)(2)(C).

[64] § 44907(d)(2)(D).

[65] § 44907(e). Invoking this action does not require that the Secretary base the determination upon TSA's airport assessment results, though an assessment may provide the basis for invoking this action.

[66] These standards and the recommended practice are reprinted with the permission of ICAO.

[67] In addition, TSA is also trying to engage in negotiations with the Philippines about providing ASSIST in that country. The current status of TSA's Capacity Development Branch's (CDB) ASSIST program in the Philippines is that TSA is waiting for the Philippine government to sign a Memorandum of Agreement for the ASSIST program. TSA created the CDB in 2007 to manage all TSA international aviation security capacity building assistance efforts, including requests for assistance in response to a host government's airport assessment results.

In: Aviation Security
Editor: Jun Bai

ISBN: 978-1-62618-048-2
© 2013 Nova Science Publishers, Inc.

Chapter 2

AVIATION SECURITY: ACTIONS NEEDED TO ADDRESS CHALLENGES AND POTENTIAL VULNERABILITIES RELATED TO SECURING INBOUND AIR CARGO[*]

United States Government Accountability Office

WHY GAO DID THIS STUDY

In 2010, passenger flights transported about 3.6 billion pounds of cargo into the United States from foreign locations. According to TSA, the introduction of explosive devices in air cargo shipments is a significant threat, and DHS was mandated to establish a system to screen 100 percent of cargo transported on all passenger aircraft traveling to, from, or within the United States by August 2010. Individuals identified as matches to the No Fly List are generally prohibited from boarding commercial aircraft because it has been determined they pose a threat to civil aviation or national security. GAO was asked to examine (1) TSA actions taken since October 2010 to enhance the security of inbound air cargo transported on both passenger aircraft and all-cargo carriers; and (2) any associated challenges TSA faces. GAO reviewed relevant requirements and documents, interviewed federal officials, and visited three airports based on cargo volume. The visits provided insights, but were

[*] This is an edited, reformatted and augmented version of United States Government Accountability Office, Publication No. GAO-12-632, dated May 2012.

not generalizable to the entire industry. This is a public version of a sensitive security report GAO issued in March 2012, which also addressed U.S. Customs and Border Protection's and TSA's use of the No Fly List to secure inbound air cargo.

WHAT GAO RECOMMENDS

GAO recommends, among other things, that DHS assess the costs and benefits of requiring all-cargo carriers to report inbound air cargo screening data. DHS concurred with GAO's recommendation and is taking actions to address it.

WHAT GAO FOUND

The Transportation Security Administration (TSA) has taken three primary actions since October 2010 to enhance security of inbound cargo on passenger and all-cargo aircraft. First, from October 2010 through May 2011, TSA issued new risk-based security requirements to focus more detailed screening measures on high-risk shipments, including prohibiting the transport of air cargo on passenger aircraft from Yemen and Somalia due to threats stemming from those areas and enhancing screening procedures for all cargo carriers. Second, the Secretary of Homeland Security established an Air Cargo Security Working Group—which included passenger and all-cargo representatives—to help identify ways to enhance the security of the air cargo system. In April 2011, the group recommended actions to enhance security such as developing mutually recognized standards for cargo screening technology, but the Department of Homeland Security (DHS) has not yet determined whether to implement them. Finally, in December 2010, DHS initiated an Air Cargo Advance Screening pilot, which is ongoing, to more readily identify high-risk cargo for additional screening prior to departing from foreign airports to the United States.

TSA has not yet met the 100 percent screening mandate as it applies to inbound air cargo due to several persistent challenges. For example, about one-third of air carriers that commented on TSA's proposal to screen all inbound cargo by the end of calendar year 2011 expressed concerns about being able to meet this date without causing significant disruptions in the air cargo supply chain. In response to these concerns, TSA proposed a new date of December 2012. TSA officials also said that it is difficult to verify the accuracy of the self-reported screening data provided by passenger air carriers used to determine the extent to which screening has been conducted in foreign countries. Further, there is no requirement for all-cargo carriers to report data comparable to passenger air carrier screening data, even though most inbound cargo is shipped into the United States by all-cargo carriers. TSA has not yet weighed the costs and benefits of requiring all-cargo carriers to submit screening data, and by doing so, TSA could determine whether this additional data could enhance its efforts to identify potential risks for inbound air cargo, develop cost effective strategies and measures to manage these risks, and provide additional assurance that all-cargo carriers are complying with TSA's enhanced screening requirements.

May 10, 2012
Congressional Requesters:

In 2010, passenger flights transported about 3.6 billion pounds of inbound air cargo—cargo arriving in the United States by air from foreign locations.[1] During the same period, all-cargo carriers—generally, aircraft configured solely for the transport of cargo (e.g., FedEx and United Parcel Service)—transported approximately 7.2 billion pounds of inbound air cargo.[2] The October 2010 discovery of explosive devices in air cargo packages on all-cargo aircraft bound for the United States from Yemen provided a vivid reminder that civil aviation remains an attractive terrorist target and highlights the continuing need to ensure that air cargo screening standards and practices effectively address emerging threats. According to the Transportation Security Administration (TSA), the security threat posed by terrorists introducing explosive devices in air cargo shipments transported on passenger and all-cargo aircraft is significant, and the risk of such an attack remains high.[3]

TSA is the federal agency responsible for securing the nation's civil aviation system, which encompasses the transport of passengers and cargo by aircraft to, from, and within the United States. TSA's responsibilities for securing air cargo include establishing security requirements governing domestic and foreign passenger air carriers that transport cargo, and overseeing the implementation of air cargo security requirements by air carriers through compliance inspections conducted by TSA transportation security inspectors. All air carriers, domestic and foreign, operating to, from, or within the United States, must maintain security measures in accordance with TSA-approved security programs and any applicable TSA-issued security directives or emergency amendments. [4] TSA is also responsible for reviewing information on all passengers travelling by commercial aircraft to, from, or within the United States against the No Fly and Selectee Lists.[5] In general, passengers identified as matches to the No Fly List are prohibited from boarding commercial flights because they present a threat to civil aviation or national security.[6] In addition to TSA, U.S. Customs and Border Protection (CBP) and foreign governments play a role in securing inbound cargo. Unlike TSA, which focuses its efforts on securing cargo prior to aircraft departures, CBP determines the admissibility of cargo to the United States and is authorized to inspect inbound air cargo for terrorists or weapons of mass destruction, or other items such as narcotics and illicit materials, upon its arrival in the United States.[7] Foreign governments may also impose separate security requirements on cargo bound for the United States from their airports.

To help enhance civil aviation security, the Implementing Recommendations of the 9/11 Commission Act of 2007 (9/11 Commission Act), enacted in August 2007, mandated that the Department of Homeland Security (DHS) establish a system within 3 years of enactment to screen 100 percent of air cargo transported on all passenger aircraft—U.S. and foreign—traveling to, from, or within the United States. [8] As of August 2010, TSA reported that it met the mandate to screen 100 percent of cargo as it applies to domestic air cargo but did not meet the mandate as it applied to inbound air cargo.[9] Although the mandate applies to both domestic and inbound cargo, TSA stated that it had to address the mandate for domestic and inbound cargo through separate systems because of limitations in its authority to regulate international air cargo industry stakeholders operating outside the United States.

In 2007 we reported on DHS's efforts to secure inbound air cargo and recommended, among other things, that TSA develop a risk-based strategy to address inbound air cargo security and clearly define TSA and CBP responsibilities for securing inbound cargo.[10] DHS concurred with but has not yet taken actions to fully address this recommendation. In 2010, we also reported on TSA's efforts to secure air cargo and recommended that TSA, among other things, develop a plan for how and when the agency intends to meet the mandate as it applies to inbound air cargo.[11] DHS concurred with the recommendation that TSA develop a plan for how and when the agency intends to meet the mandate as it applies to inbound cargo. In late March 2012, TSA provided a tactical plan including, among other things, a timeline that would require air carriers to screen 100 percent of inbound air cargo transported on passenger aircraft by December 1, 2012. TSA's plan to implement the screening mandate is a key step in its efforts to secure inbound air cargo and it will be important for the agency to continue to monitor passenger air carrier efforts to adhere to TSA's plan for meeting the mandate. In 2010, we also recommended that TSA establish a mechanism to verify data on screening conducted on inbound air cargo.[12] DHS concurred in part with our recommendation that the agency develop a mechanism to obtain and verify the accuracy of screening data and TSA took steps to obtain screening data. We discuss the actions TSA has taken to verify screening data since 2010 later in this report. Finally, in October 2011, we reported on TSA's efforts to determine whether foreign airports that provide service to the United States are maintaining and carrying out effective security measures, including those related to air cargo.[13] We recommended, among other things, that TSA develop a mechanism to evaluate the results of completed assessment activities to determine any trends and target future activities and resources. DHS concurred with our recommendation and stated TSA is taking steps to address it.

The DHS Appropriations Act, 2012, requires the Administrator of TSA to submit to the Committees on Appropriations of the Senate and the House of Representatives a report that either (1) certifies that the requirement for screening all air cargo on passenger aircraft has been met or (2) includes a strategy to comply with the screening requirement that includes a plan for meeting the screening requirement and that specifies the percentage of such air cargo that is being screened and the schedule for achieving screening of 100 percent of such air cargo.[14] The Administrator is to submit such a report not later than 180 days after enactment (enacted December 23, 2011) and every 180 days thereafter until the Administrator certifies that TSA has achieved the 100 percent screening mandate.

You asked us to review DHS's progress and challenges in screening and securing air cargo transported to the United States. In response to this request, this report addresses the following questions:

1. What actions has TSA taken since October 2010 to enhance the security of inbound air cargo transported on both passenger and all-cargo aircraft?
2. What challenges, if any, has TSA faced in enhancing the security of inbound air cargo?

This report is a public version of a prior sensitive report that we provided to you in March 2012. DHS deemed some of the information in the prior report sensitive security information, which must be protected from public disclosure.[15] Therefore, this report omits sensitive information regarding a third question about how agencies secure cargo using the No Fly List. In addition, at DHS's request, we have omitted information on a potential vulnerability we identified related to the No Fly List and additional steps CBP and TSA could take to strengthen how the agencies use the No Fly List to secure inbound air cargo. The information provided in this report is more limited in scope as it excludes such sensitive information, but the overall methodology used for both reports is the same.

To answer the first and second objectives, we reviewed and analyzed TSA's air cargo security policies and procedures, screening program documents, and security directives and emergency amendments issued in response to the October 2010 bomb attempt, as well as industry comments to TSA's proposed screening program changes and date for implementing the mandate as it applies to inbound cargo. We assessed TSA's efforts against DHS risk management criteria and Standards for Internal Control in the Federal Government.[16] We also interviewed TSA and CBP air cargo program officials to discuss progress and challenges in securing inbound air cargo, including agency efforts taken in response to the October 2010 bomb attempt. In addition, we conducted site visits to three airports in the United States, selected because they are within the top five commercial airports that handle the greatest cargo volume in the United States. Although our site visits were based on a nonprobability sample and cannot be generalized to the entire air cargo industry, this sample allowed us to observe domestic cargo screening operations and programs in various parts of the country with differing air cargo volumes and commodities, which also provided insights on inbound cargo screening operations. We did not observe CBP inbound air cargo screening operations as part of our site visits because we focused on the agency's efforts to target elevated-risk inbound air shipments.

We also interviewed TSA air cargo program officials and representatives from six domestic and foreign passenger air carriers to obtain their views on TSA's efforts to implement the screening mandate including TSA's proposed security program changes. We selected the six passenger air carriers to obtain a representation of air carriers serving different international regions. Moreover, these six air carriers are among the passenger air carriers that transported the greatest volume of international cargo by weight in 2010, including cargo transported inbound to the United States. We also interviewed two of the largest passenger air carrier industry associations representing foreign and domestic air carriers to obtain their views on TSA's security program changes for meeting the screening mandate as it applies to inbound air cargo. Their views cannot be generalized to all foreign and domestic air carriers but provided insights. For all-cargo carriers, we interviewed representatives of four all-cargo air carriers, three of which we interviewed at the three airports we visited, to obtain their views on, among other things, TSA's efforts to enhance inbound air cargo security in response to the October 2010 bomb attempt. We selected these four all-cargo carriers to obtain a representation of all-cargo carriers that transport varying volumes of international cargo by weight, and two of the four all-cargo carriers we interviewed are responsible for transporting approximately 25 percent of inbound air cargo. Because we selected a nonprobability sample of all-cargo carriers, the results of these interviews cannot be generalized to other all-cargo carriers. However, this sample allowed us to understand all-cargo operations and provided important perspective on all-cargo air carrier efforts to secure inbound cargo. In addition, we interviewed two industry associations that represent all-cargo carriers, including the two largest international all-cargo carriers.

We conducted this performance audit from February 2011 through May 2012 in accordance with generally accepted government auditing standards. Those standards require that we plan and perform the audit to obtain sufficient, appropriate evidence to provide a reasonable basis for our findings and conclusions based on our audit objectives. We believe that the evidence obtained provides a reasonable basis for our findings and conclusions based on our audit objectives.

BACKGROUND

Roughly 10 billion pounds of cargo are transported on inbound flights to the United States per year. Approximately 76 passenger air carriers and 70 all-cargo carriers transport cargo to the United States and approximately 300 foreign airports in 100 countries provide last point of departure flights to the United States.[17] This cargo ranges in size from 1 pound to several tons and ranges in type from perishable commodities to machinery. Air cargo can include such varied items as electronic equipment, automobile parts, clothing, medical supplies, fresh produce, and human remains. As seen in figure 1, cargo can be shipped in various forms, including unit load devices (ULD) that allow many packages to be consolidated into one large container or pallet that can be loaded onto an aircraft; wooden skids or crates; as well as individually wrapped/boxed pieces known as loose or break bulk cargo.

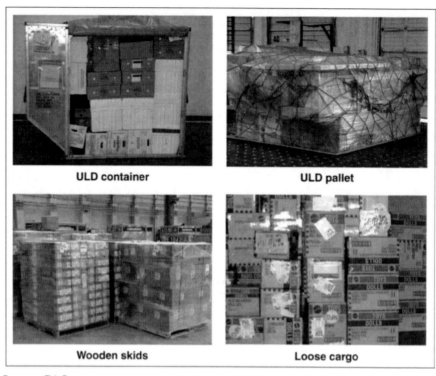

ULD container

ULD pallet

Wooden skids

Loose cargo

Source: GAO.

Figure 1. Various Means of Shipping Air Cargo.

Participants in the international air cargo shipping process can include shippers, such as individuals or manufacturers; freight forwarders or regulated agents, who consolidate shipments and deliver them to air carriers; air carrier cargo handling agents, who process and load cargo onto aircraft on behalf of air carriers; and passenger and all-cargo carriers that store, load, and transport air cargo. Figure 2 depicts the two primary ways a shipper may send inbound air cargo to the United States.

The Aviation and Transportation Security Act (ATSA), enacted into law shortly after the September 11, 2001, terrorist attacks, established TSA and gave the agency responsibility for securing all modes of transportation, including the nation's civil aviation system, which includes air carrier operations (domestic and foreign) to, from, and within the United States.[18]

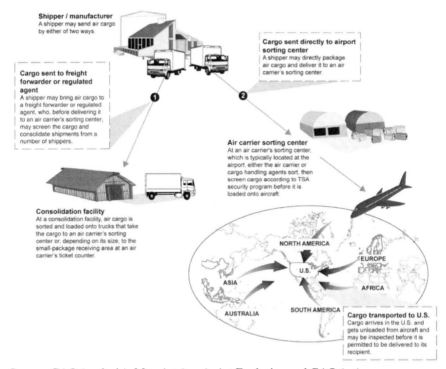

Source: GAO (analysis); Map Art (map); Art Explosion and GAO (art).

Figure 2. Flow of Inbound Air Cargo Transported to the United States.

Among other things, ATSA requires that TSA provide for the screening of all passengers and property, including cargo, transported on passenger aircraft originating in the United States.[19] ATSA further requires that a system be in operation, as soon as practicable, to screen, inspect, or otherwise ensure the security of the cargo transported by all-cargo aircraft—generally, aircraft configured solely for the transport of cargo—to, from, and within the United States but did not establish a firm deadline for the implementation of such a system.[20] Consistent with ATSA and subsequent statutory requirements, including the 9/11 Commission Act, TSA has focused its efforts on establishing a system to secure cargo on passenger aircraft. Following the October 2010 bomb attempt originating in Yemen, TSA expedited efforts to further secure cargo on all-cargo aircraft inbound to the United States to include mitigating threats posed by the introduction of an improvised explosive devices in cargo, in addition to the original focus on detecting and preventing stowaways intent on hijacking the aircraft and using it as a weapon.

The 9/11 Commission Act defines screening for purposes of the air cargo screening mandate as a physical examination or the use of nonintrusive methods to assess whether cargo poses a threat to transportation security. [21] Specific screening methods outlined in the 9/11 Commission Act include X-ray systems, explosives detection systems (EDS), explosives trace detection (ETD), explosives detection canine teams certified by TSA, physical search together with manifest verification, and additional methods approved by the TSA Administrator.[22] However, solely performing a review of information about the contents of cargo or verifying the identity of the cargo's shipper does not constitute screening for purposes satisfying the mandate. Pursuant to the mandate, the system for screening air cargo must provide a level of security commensurate with the level of security for the screening of checked baggage.[23]

TSA's Transportation Sector Network Management (TSNM) Air Cargo Division is responsible for developing air cargo regulations, establishing program regulations for the development of technological solutions, and policies that enhance the security of the air cargo supply chain while maintaining TSA's commitment to ensure the flow of commerce. In fiscal year 2011, TSA's budget for air cargo programs was approximately $115 million, with $26 million dedicated to domestic and inbound air cargo security efforts, $74 million dedicated to air cargo inspectors and proprietary canines, and $15 million dedicated to the National Explosives Detection Canine Team Program. Approximately $124 million was made available to TSA to support its air cargo programs for fiscal year 2012.[24]

Risk management is a tool for informing policymakers' decisions about assessing risks, allocating resources, and taking actions under conditions of uncertainty. We have reported that risk management entails a continuous process of managing risk through a series of actions, including setting strategic goals and objectives, assessing risk, evaluating alternatives, selecting initiatives to undertake, and implementing and monitoring those initiatives. We have previously reported that a risk management approach can help to prioritize the programs designed to combat terrorism. Risk management, as applied in the transportation security context, can help federal decision makers determine where and how to invest limited resources within and among the various modes of transportation.[25] The DHS National Infrastructure Protection Plan (NIPP) includes a risk management framework that consists of six steps, which closely reflects GAO's risk management framework.[26] Like GAO's framework, the NIPP's risk management framework is a repetitive process that continuously uses the results of each step to inform the activities in both subsequent and previous steps over time. The NIPP risk management framework is designed to produce a systematic and comprehensive understanding of risk and ultimately provide for security investments based on this knowledge of risk.

TSA HAS TAKEN ACTIONS TO ENHANCE THE SECURITY OF CARGO ON INBOUND AIRCRAFT SINCE OCTOBER 2010

TSA has taken three primary actions to enhance security of inbound cargo on passenger and all-cargo aircraft following the October 2010 bomb attempt originating in Yemen.

TSA issued new screening requirements aimed at enhancing the security of cargo on passenger and all-cargo aircraft. Beginning in October 2010 through May 2011, TSA required new risk-based security procedures aimed at focusing more detailed screening measures on high-risk shipments. These security measures included, among other things, prohibiting the transport of air cargo on passenger aircraft from Yemen and Somalia due to threats stemming from those areas.[27]

Moreover, the Yemen incident raised questions about the effectiveness of technology and screening protocols used to screen air cargo because suspected packages were screened multiple times, using multiple methods, at various locations yet the threat items were detected only after foreign law enforcement officials opened the shipments based on a tip from an intelligence source. According to TSA, the threat item used in the incident likely would not have been detected by air carriers using TSA screening protocols in place at that time because screening requirements for all-cargo carriers focused on preventing and detecting stowaways or contraband items and not on detecting explosive devices and, for passenger air carriers, screening requirements primarily focused on detecting assembled explosive devices rather than on the types of specific components used to construct the explosive device associated with the Yemen incident. Following the October 2010 bomb attempt, DHS simulated the conditions of the incident in a laboratory setting, and TSA found that existing approved technologies are capable of detecting the materials used in the bomb attempt when used by screeners trained in protocols for identifying those specific materials. Subsequent to these tests, and as noted above, TSA implemented new risk-based security procedures to enhance the effectiveness of TSA screening requirements.

In addition, officials from all four all-cargo carriers we spoke with stated that in some cases they implement their own inbound air cargo security measures, in addition to those required by TSA requirements, to safeguard their staff, cargo shipments, and aircraft. Officials from three of the six passenger air carriers we spoke with also provided examples of additional security measures they implement.[28]

DHS instituted working groups with air cargo industry stakeholders to identify ways to enhance air cargo security. Following the October 2010 bomb attempt originating in Yemen, in January 2011, the Secretary of Homeland Security established an Air Cargo Security Working Group, composed of four subgroups, to obtain advice and consultations from air cargo security stakeholders—including passenger and all-cargo representatives—on ways to enhance the security of the air cargo system.[29] The Air Cargo Security Working Group briefed the Secretary of Homeland Security, the Commissioner for CBP, and the TSA Administrator in April 2011 on proposed solutions, and recommended that TSA reevaluate the agency's implementation plan, timeline, and resources related to the National Cargo Security Program (NCSP) recognition program, a TSA effort to recognize the security programs of foreign countries. Other proposed solutions included establishing an international trusted shipper program to perform cargo security and screening throughout the international supply chain, before its transport from last point of departure airports, and developing mutually recognized standards for cargo screening technology. Given that DHS is currently considering actions to take in response to the recommendations and which ones to implement, it could not provide us with specifics on the implementation of the Air Cargo Security Working Group recommendations. According to DHS, the Air Cargo Security Working Group may meet again in 2012, to discuss its role in supporting the implementation of the recommendations but a date has not yet been established. Following this meeting, the members plan to work on a standalone basis through the four subgroups and advise the Secretary of Homeland Security upon her request.

DHS initiated an Air Cargo Advance Screening (ACAS) pilot. DHS initiated this pilot in December 2010 to more readily identify high-risk cargo for additional screening prior to all-cargo and passenger aircraft departing from foreign airports to the United States. The aim of the pilot, which is ongoing, is to determine whether it is feasible for air carriers to submit air cargo manifest data to CBP prior to departure from all foreign last point of departure airports to allow CBP to analyze, target, and, if needed, issue instructions to air carriers to provide additional cargo information or take additional security measures before such cargo is loaded onto aircraft. DHS initially focused on all-cargo express carriers and companies due to the elevated risk highlighted by the October 2010 incident.[30]

As of December 2011, the pilot included select airports from a number of countries and regions. Under current CBP requirements, CBP must receive manifest data for air cargo shipments from air carriers no later than 4 hours prior to the flight's arrival in the United States or no later than the time of departure ("wheels up" and en route directly to the United States) from locations in North America, including Canada, Mexico, the Caribbean, Central America, and parts of South America north of the Equator.[31]

Under the pilot program, however, participants provide manifest data prior to loading cargo aboard aircraft.

While the pilot initially focused on four all-cargo express carriers, the agencies began expanding the pilot to include passenger air carriers in the fall of 2011. Four passenger carriers have committed to joining the pilot. According to DHS officials, early results demonstrate that all-cargo express carriers and companies are able to submit manifest data, in some cases, a number of hours in advance of existing CBP requirements for receipt of manifest data described above.

However, DHS officials stated that it will be difficult to obtain manifest data far in advance as the pilot expands to include additional carriers, such as passenger air carriers, and other geographic locations due to the need to develop additional protocols with entities (carriers and freight forwarders) that may not have previously transmitted data to DHS. Further, five of six passenger air carriers we spoke with stated that submitting this information far in advance will be challenging because passenger air carriers typically receive manifest data from freight forwarders just hours before departure. To address these challenges, DHS officials explained that they are meeting with freight forwarders—which have shipment information before air carriers—to discuss the feasibility of having forwarders submit manifest information directly to CBP as part of the ACAS pilot.

Because these efforts are in the initial stages, it is too early to determine the extent to which CBP and TSA will be able to obtain advance cargo manifest data prior to departure from passenger air carriers and its impact on strengthening security.

CHALLENGES PERSIST IN MEETING SCREENING MANDATE, AND ADDITIONAL EFFORTS COULD FACILITATE OVERSIGHT OF CARRIERS' ALL-CARGO SCREENING EFFORTS

Challenges Exist In Meeting Screening Requirements for Inbound Cargo Transported on Passenger and All-Cargo Aircraft

Air carriers and TSA face challenges that, among other things, could limit TSA's ability to meet the mandate to screen 100 percent of inbound air cargo transported on passenger aircraft and to provide reasonable assurance that screening is being conducted at reported levels. All cargo carriers subject to TSA regulation also reported facing challenges in implementing new TSA screening requirements implemented after the October 2010 Yemen incident.

Passenger Air Carriers Reported Logistical Challenges Implementing Proposed Screening Requirements. In January 2011, TSA proposed additional changes to passenger aircraft security requirements outlined in the Aircraft Operator Standard Security Program (AOSSP) and the Model Security Program (MSP) to further enhance the security of air cargo departing foreign locations by requiring 100 percent screening of cargo previously exempt from screening. Current TSA requirements call for air carriers to screen a certain percentage of all cargo. [32] TSA's proposed changes were to go into effect on December 31, 2011, and would have required passenger air carriers to screen 100 percent of cargo. TSA proposed these changes as part of the agency's efforts to meet the 9/11 Commission Act mandate for 100 percent screening of inbound air cargo transported on passenger aircraft by the end of calendar year 2011—2 years earlier than the TSA administrator reported to Congress in November 2010.

In commenting on TSA's proposal to achieve 100 percent screening of inbound cargo by the end of calendar year 2011, passenger air carriers expressed concerns about being able to meet the 100 percent screening mandate as it applies to inbound cargo by December 31, 2011, stating that it would cause significant disruptions in the air cargo supply chain by requiring, among other things, 100 percent screening, including screening of cargo previously exempt from screening. In comments submitted to TSA, 6 of 19 foreign air carriers and one industry association were opposed to the new proposed date due to insufficient time to implement the proposed requirements and potential detrimental impact to air carrier operations in the form of higher costs and slower screening times. Five of these 6 foreign air carriers are among the passenger air carriers that transport the greatest volume of international cargo by weight, and the association represents hundreds of air carriers whose flights comprise over 80 percent of international air traffic. In addition, 5 of 19 foreign airlines and the 2 industry associations representing foreign airlines reported that they opposed TSA's proposed changes to screen 100 percent of cargo previously exempt from screening, stating that this would delay the processing of cargo and potentially disrupt the flow of commerce. Moreover, 3 of the 6 passenger air carriers we spoke with stated that they would no longer be able to transport certain types of cargo if TSA were to require 100 percent screening of cargo previously exempt from screening.

Two of the six passenger air carriers we spoke with expressed concern about new requirements calling for the screening of all inbound air at the last point of departure because it would cause disruptions in the chain of custody throughout the air cargo supply chain without providing any additional security benefit. Specifically, they stated that some inbound air cargo is already screened at points of origin and that the new requirements would in effect result in rescreening of already screened cargo. Additionally, one of six passenger air carriers and one industry association—representing U.S. air carriers that transport more than 90 percent of all U.S. airline and passenger cargo traffic—commented on the challenges in screening ULD pallets or containers, which are the primary means of transporting air cargo on wide-body passenger aircraft on both domestic and inbound aircraft. Since TSA has not approved or qualified any equipment to screen cargo transported on ULD pallets or containers, these air carriers and associations commented that breaking down ULDs in order to rescreen cargo would cause further delays and inefficiencies.[33] For example, three foreign air carriers commented that it is highly impractical to unload screened cargo from ULDs, rescreen that cargo, and then reload the ULDs and that doing so would reduce operational efficiencies, result in delays, and expose cargo to damage and theft.

In addition, TSA reported that a number of passenger air carriers informed the agency that they would have difficulty implementing additional security measures outlined in the security directives and emergency amendments to the AOSSP and MSP. [34] According to TSA officials, air carriers request alternative security measures when they are unable to comply with the exact language contained in the TSA-issued security program, but can meet the intent of the requirement through an alternative security measure, which must result in an equal or better security outcome. [35] These passenger carriers—both domestic and foreign—have requested alternative security measures to implement the security directives and emergency amendments, which TSA has approved.

To address these issues an industry group and one air carrier suggested TSA establish/recognize international supply chain security programs akin to the Certified Cargo Screening Program (CCSP), which is a voluntary cargo screening program for air cargo industry participants that allows screening to take place at various points along the air cargo supply chain in the United States. [36] This suggestion is similar to the one put forth by the DHS working group made up of air cargo industry stakeholders which proposed establishing an international trusted shipper program to perform cargo security and screening throughout the international supply chain, before its transport from last point of departure airports. These stakeholders contend that such a program could assist air carriers in implementing the 100 percent screening mandate as was done in the United States under the CCSP.

According to TSA officials the agency plans to respond to the challenges raised by industry stakeholders by revising the proposed requirements, which may include, among other things, risk-based security procedures similar to those TSA issued following the October 2010 bomb attempt originating in Yemen. TSA plans to publish the revised security requirements in early May 2012. TSA anticipates that the new proposed requirements will become effective in June 2012. In commenting on the sensitive security version of our report, DHS stated that TSA finalized a tactical plan, with a proposed timeline, to achieve 100 percent screening of inbound air cargo transported on passenger aircraft. In late March 2012, TSA provided us with the tactical plan that included, among other things, a timeline that would require air carriers to screen 100 percent of inbound air cargo transported on passenger aircraft by December 1, 2012. TSA's plan to implement the screening mandate is a key step in its efforts to secure inbound air cargo and it will be important for the agency to continue to monitor passenger air carriers' efforts to adhere to TSA's plan for meeting the mandate.

TSA Faces Challenges Verifying Screening Data on Inbound Passenger Cargo. TSA relies on data submitted to the agency by air carriers to determine the amount of inbound air cargo screened in accordance with TSA screening requirements. As noted above, in January 2011, TSA announced that it was planning to meet the 9/11 Commission Act mandate as it applies to inbound air cargo transported on passenger aircraft by the end of calendar year 2011. TSA officials determined that it was feasible to accelerate its plans at that time based on screening data reported by air carriers in late

2010 and air carrier feedback. TSA officials subsequently determined in the spring of 2011 that air carriers were screening cargo at lower levels than previously reported due to industry stakeholder confusion regarding TSA's proposed requirements for screening inbound air cargo. As of September 2011, TSA officials stated that current air carrier-reported screening percentages—which they estimate to be about 80 percent—are based on actual data reported by air carriers, but agreed that it is difficult to verify the accuracy of the screening data reported by air carriers with reasonable assurance. As of December 2011, the air carrier data have not been independently verified for accuracy since TSA has not developed a mechanism to cross-reference local screening logs—which include screening information on specific shipments—with screening reports submitted by air carriers to TSA that do not contain such information. In addition, TSA counts all inbound air cargo from foreign countries that require 100 percent screening and that are recognized under the NCSP as screened for the purposes of meeting the mandate. However, while TSA officials stated that they discuss screening percentages with foreign government officials, the agency does not have a mechanism to conduct additional data verification to assess whether screening is conducted at, above, or below the required levels.

In order to more accurately identify the level of screening being conducted on inbound air cargo, we recommended in June 2010 that TSA develop a mechanism to verify the accuracy of all screening data through random checks or other practical means.[37] TSA concurred in part with our recommendation and issued changes to air carriers' standard security programs that require air carriers to report inbound cargo screening data to TSA. TSA officials told us that in May 2010 the agency created a reporting requirement for air carriers to provide screening data on a monthly basis. TSA also stated that inspectors review screening data, among other things, when inspecting air carriers as part of the agency's air carrier compliance inspections. However, since TSA still has not developed a mechanism to verify the accuracy of the data reported by air carriers, the agency has not yet fully met the intention of our recommendation. It will be important for TSA to continue to work towards ensuring verification of inbound air cargo screening data submitted by air carriers and that inbound air cargo is screened in accordance with the mandate.

Harmonizing U.S. and Foreign Air Cargo Security Standards Is Difficult. TSA has developed an NCSP recognition program that would review and recognize the air cargo security programs of foreign countries if TSA deems those programs as providing a level of security commensurate with TSA's air cargo security standards. As of November 2011, TSA has recognized 2 countries as providing a level of security commensurate with current U.S. air cargo security standards, and continues to evaluate the comparability of air cargo security programs for several other countries. TSA also recognizes 2 additional countries under the legacy program that preceded the NCSP recognition program and the agency is evaluating these countries against the current NCSP standards for renewal. TSA officials stated that the NCSP recognition program is a key effort in meeting the 100 percent screening mandate because it will eliminate the need for air carriers to comply with two country's security programs. However, according to TSA officials, evaluating foreign countries' air cargo security programs is challenging because the agency is dependent upon countries' willingness and ability to work with TSA officials to ensure their programs are commensurate to those of the United States. As of November 2011, TSA was coordinating with 16 countries, including 12 of the top 20 passenger air cargo volume countries, which according to TSA officials and based on BTS data, represent 51 percent of the inbound air cargo volume by weight to the United States.

We previously reported that given the challenges TSA faces in meeting the screening mandate for inbound air cargo, the agency would be better positioned to meet the mandate by developing a plan to account for these challenges. As such, in June 2010, we recommended that TSA develop a plan, with milestones, for how and when the agency intends to meet the mandate as it applies to inbound cargo. TSA concurred with the recommendation and stated that the agency was drafting milestones as part of a plan to generally require air carriers to conduct 100 percent screening by a specific date. TSA has not yet fully implemented this recommendation but agency officials stated that it plans to do so by December 2012. In light of the number of challenges facing TSA in its efforts to address the screening mandate for inbound air cargo, we continue to believe that our recommendation has merit and that it will be important for TSA to document such a plan.

ALL-CARGO CARRIERS REPORTED CHALLENGES MEETING NEW SCREENING REQUIREMENTS

A number of foreign and domestic all-cargo carriers subject to TSA regulation initially notified the agency that they would have difficulty in implementing the new TSA screening requirements introduced after the October 2010 incident, including a requirement to screen all cargo using a certain type of screening technology, because all-cargo carriers stated that they do not have such technology at all locations.[38] Moreover, a number of domestic all-cargo carriers requested TSA approval to implement alternative security procedures (as an alternative to requirements in security directives and emergency amendments) since they stated that it would be challenging to implement these new security measures. As previously discussed, according to TSA officials, air carriers request alternative procedures when they are unable to comply with the exact language contained in the requirement, but can meet the intent of the requirement through an alternative procedure, which must result in an equal or better security outcome. TSA approved all of these requests, as of December 2011. For example, some all-cargo carriers received TSA approval to use different procedures or different models— or versions— of screening equipment to screen cargo in lieu of using equipment on TSA's list of approved screening equipment. Officials from all-cargo carriers explained that they only deploy screening technology to airports that they deem to be high risk and consequently did not have TSA-approved screening technology available for use at other last point of departure foreign airports that they deem lower risk.

Reporting Screening Data Could Facilitate Oversight of All-Cargo Carriers' Compliance with Security Requirements

TSA relies on data submitted by passenger carriers to determine the amount of air cargo screened on inbound passenger aircraft but there is no requirement for all-cargo carriers to report comparable screening data to TSA even though most of the cargo shipped from abroad into the United States is shipped on all-cargo carriers. Thus, TSA does not know the extent to which all-cargo carriers are screening cargo or meeting the enhanced screening requirements introduced after the October 2010 incident in Yemen. In 2010, all-cargo carriers transported approximately 67 percent (7.2 billion pounds) of the total cargo (10.8 billion pounds) transported to the United States.[39] According to officials from two global all-cargo carriers we spoke with, responsible for transporting approximately 25 percent of inbound air cargo on all-cargo flights, submitting such information to TSA would be feasible because they are already collecting this data internally. However, officials from the other two all-cargo carriers we interviewed stated that reporting screening data to TSA would be challenging because of staffing limitations at foreign airports necessary to compile and submit the screening data or because such data may not be available.

TSA officials explained that the agency does not currently require that all-cargo carriers submit screening data to demonstrate compliance with enhanced screening requirements because the agency has focused its efforts on collecting data from passenger air carriers in support of meeting the 100 percent mandate. In addition, TSA officials stated that the agency is attempting to gain some visibility over all-cargo carrier operations through the ACAS pilot.[40] TSA officials stated that as part of the pilot program's future expansion, the agency may consider opportunities to capture additional inbound air cargo information, but has not yet weighed the costs and benefits of requiring such all-cargo carriers to submit data on their screening efforts because the agency has focused its efforts on establishing the pilot program. The pilot program is a key effort to identify high-risk cargo prior to aircraft departing from foreign airports, but the pilot is not intended to provide TSA with screening data, which if collected and verified, could provide additional assurance that all-cargo carriers are complying with the agency's enhanced screening requirements.

DHS's risk management policy directs its components to adopt risk management practices in order to, among other things, inform decisions to enhance security and manage homeland security risks. [41] As part of this policy, DHS components are to incorporate risk management processes to identify potential risks, and develop and analyze alternative strategies to manage risks considering projected costs, benefits, and ramifications of each alternative to manage or mitigate the risk. By assessing the costs and benefits of requiring all-cargo carriers to submit screening data to TSA in accordance with DHS's policy on risk management, the agency could determine whether these additional data could enhance the agency's efforts to identify potential risks to inbound air cargo and develop cost effective strategies and measures to manage these risks. In addition, Standards for Internal Control in the Federal Government underscore the need for developing effective controls for meeting program objectives and complying with applicable regulation and capturing information needed to meet program objectives.[42] These standards also include designing controls to assure that ongoing monitoring occurs in the course of normal operations and determining that relevant, reliable, and timely information is available for management decision-making purposes. TSA could better determine what actions are needed, if any, to ensure that all-cargo carriers are complying with the agency's enhanced screening requirements by assessing the costs and benefits of requiring all-cargo carriers to report data on screening conducted.

CONCLUSION

DHS and TSA have taken steps to better secure inbound air cargo in the wake of the October 2010 bomb attempt in Yemen. However, TSA continues to face a number of challenges that complicate its efforts to meet the 9/11 Commission Act screening mandate as it relates to inbound air cargo transported on passenger aircraft. These challenges influenced the agency's decision to propose a new date for screening 100 percent of inbound cargo on passenger aircraft of December 2012. Moreover, while TSA has taken steps to enhance cargo screening on all cargo carriers, TSA does not have full visibility over their efforts to screen air cargo since it does not require all-cargo air carriers to report any data on their screening efforts. Assessing the costs and benefits of requiring all-cargo carriers to submit screening data could help TSA determine whether these additional data could enhance the agency's efforts to identify potential risks to inbound air cargo and develop cost-effective strategies and measures to manage these risks.

RECOMMENDATION FOR EXECUTIVE ACTION

To help DHS address challenges in meeting the air cargo screening mandate as it applies to inbound air cargo, mitigate potential air cargo security vulnerabilities, and enhance overall efforts to screen and secure inbound air cargo, we recommend that the Secretary of Homeland Security direct the Administrator of TSA to assess the costs and benefits of requiring all-cargo carriers to report data on the amount of inbound air cargo screening being conducted.

AGENCY COMMENTS AND OUR EVALUATION

We provided a draft of this report to DHS, TSA, and CBP for review and comment. In commenting on our report, DHS stated that it concurred with the recommendation and identified actions taken or planned to implement it.

DHS concurred with the recommendation that TSA assess the costs and benefits of requiring all-cargo carriers to report data on the amount of inbound air cargo screening being conducted. DHS stated that as part of the ACAS pilot, TSA is taking steps to require carriers to provide confirmation that screening has been conducted in accordance with TSA's enhanced screening requirements. DHS also stated that when fully implemented, the ACAS pilot will provide the capability to report high-risk inbound air cargo shipments screened by all-cargo carriers. We support TSA's actions to identify and report on high-risk cargo being screened using ACAS capabilities. Given that the ACAS pilot is currently limited to a relatively small number of participants, expanding the pilot to include a greater number of all-cargo carriers that transport air cargo to the United States could provide TSA with additional assurance that all-cargo carriers are complying with the agency's enhanced screening requirements. Such action, when fully implemented, would further address the intent of our recommendation.

In its comments, DHS also referred to a second recommendation to the Administrator of TSA and Commissioner of CBP related to how the agencies use the No Fly List to secure inbound air cargo. Because DHS deemed the details of this recommendation and its response as sensitive security information, they are not included in the public version of the report.

Finally, in commenting on our report, DHS provided an update on TSA's efforts to implement the mandate to screen 100 percent of inbound air cargo transported on passenger aircraft. DHS stated that TSA has finalized a tactical plan, with a proposed timeline, to achieve 100 percent screening of inbound air cargo transported on passenger aircraft. The plan involves, among other things, a timeline for implementing a revised security program by summer 2012 that would require air carriers to screen 100 percent of inbound air cargo transported on passenger aircraft by December 1, 2012. The plan also involves use of the ACAS pilot by passenger air carriers to submit data and proposed amendments to TSA's carrier security requirements. TSA's plan to implement the screening mandate is a key step to securing inbound air cargo and it will be important for the agency to continue to monitor its efforts to ensure that passenger air carriers adhere to TSA's plan for meeting the mandate.

DHS and the Federal Bureau of Investigation's Terrorist Screening Center also provided us with technical comments, which we incorporated as appropriate.

Stephen M. Lord
Director, Homeland Security and Justice Issues

End Notes

[1] Based on 2010 Bureau of Transportation Statistics (BTS) data.

[2] Based on 2010 BTS data. All-cargo aircraft are aircraft configured solely for the transport of cargo.

[3] Prior to the Yemen incident, TSA considered the primary threat to all-cargo aircraft to be stowaways hijacking the aircraft and using it as a missile to cause mass destruction.

[4] See 49 U.S.C. §§ 44903(c), 44906. See also 49 C.F.R. §§ 1544.3, 1544.101-1544.105, 1544.305, 1546.3, 1546.101-1546.105.

[5] The No Fly and Selectee Lists are subsets of the consolidated terrorist watchlist maintained by the Federal Bureau of Investigation's Terrorist Screening Center that contains the names of individuals with known or suspected links to terrorism.

[6] In general, persons who are deemed to be a threat to civil aviation or national security but do not meet the criteria of the No Fly List are placed on the Selectee List and are to receive additional screening prior to being permitted to board an aircraft.

[7] A weapon of mass destruction could include nuclear, biological, chemical, or radiological devices.

[8] See Pub. L. No. 110-53, § 1602(a), 121 Stat. 266, 477-79 (2007) (codified at 49 U.S.C. § 44901(g)) (providing that the system to screen 100 percent of all cargo transported by passenger aircraft shall be established no later than 3 years after enactment).

[9] For the purposes of this report, domestic cargo refers to cargo transported by air within the United States and from the United States to a foreign location by both U.S. and foreign air carriers.

[10] GAO, *Aviation Security: Federal Efforts to Secure U.S.-Bound Air Cargo Are in the Early Stages and Could Be Strengthened,* GAO-07-660 (Washington, D.C.: Apr. 30, 2007).

[11] GAO, *Aviation Security: TSA Has Made Progress but Faces Challenges in Meeting the Statutory Mandate for Screening Air Cargo on Passenger Aircraft,* GAO-10-446 (Washington, D.C.: June 28, 2010).

[12] GAO-10-446.

[13] GAO, *Aviation Security: TSA Has Taken Steps to Enhance Its Foreign Airport Assessments, but Opportunities Exist to Strengthen the Program,* GAO-12-163 (Washington, D.C.: Oct. 21, 2011).

[14] See Pub. L. No. 112-74, Div. D, § 548, 125 Stat. 786, 977-78 (2011).

[15] See 49 C.F.R. pt. 1520.

[16] GAO, *Standards for Internal Control in the Federal Government,* GAO/AIMD-00-21.3.1 (Washington, D.C.: November 1999).

[17] Approximately 17 foreign airports with nonstop flights to the United States (known as last points of departure) serve only all-cargo carriers.

[18] See generally Pub. L. No. 107-71, 115 Stat. 597 (2001).

[19] See Pub. L. No. 107-71, § 110(b), 115 Stat. at 614-16 (codified as amended at 49 U.S.C. § 44901).

[20] See 49 U.S.C. § 44901(f).

[21] See 49 U.S.C. § 44901(g)(5).

[22] An EDS machine uses computed tomography technology to automatically measure the physical characteristics of objects in baggage. The system automatically triggers an alarm when objects that exhibit the physical characteristics of explosives are detected. An ETD machine is used to chemically analyze trace materials after a human operator swabs checked baggage to identify any traces of explosive material.

[23] For example, TSA requires that all checked baggage transported on passenger aircraft from a U.S. airport be screened using an explosives detection system or by explosives trace detection, and air carrier security programs address baggage screening requirements for aircraft bound for the United States from foreign locations. See 49 U.S.C. §§ 44901(d), (g)(2). See also 49 C.F.R. §§ 1544.203, 1544.213, 1546.203.

[24] See H.R. Rep. No. 112-331, at 969-70, 72 (Dec. 15, 2011) (conference report).

[25] For information on DHS's and TSA's risk management framework, see GAO, *Transportation Security: Comprehensive Risk Assessments and Stronger Internal Controls Needed to Help Inform TSA Resource Allocation,* GAO-09-492 (Washington, D.C.: Mar. 27, 2009).

[26] The six steps contained in the NIPP are (1) Set security goals; (2) Identify assets, systems, networks, and functions; (3) Assess risks (consequences, vulnerabilities, and threats); (4) Prioritize; (5) Implement protective programs; and (6) Measure effectiveness.

[27] Requirements also apply to all-cargo carriers.

[28] DHS deemed details on specific measures taken by air carriers as sensitive security information. Thus, they are not included in this report.

[29] DHS's Air Cargo Security Working Group consists of four subgroups: (1) Information subgroup, whose objective is to, among other things, enhance intelligence and information sharing among federal stakeholders and between the U.S. government and private sector entities; (2) technology and capacity building subgroup, whose objective is to review technology standards and develop suggestions for addressing technology limitations; (3) global cargo programs subgroup, whose objective is to review and explore opportunities for enhanced public-private coordination as DHS works to address statutory requirements for screening 100 percent of inbound air cargo; and (4) global mail subgroup, whose objective is to, among other things, identify potential vulnerabilities for global mail and propose alternative processes and procedures to ensure the safety of mail transported by air.

[30] All-cargo express carriers and companies focus on transporting cargo under quick time frames.

[31] See 19 C.F.R. § 122.48a(b).

[32] Details on TSA's screening requirements are deemed sensitive security information and not included in this report.

[33] As we reported in June 2010, about 96 percent of inbound air cargo is transported to the United States on wide-body aircraft. TSA officials stated that they continue to work with technology vendors to identify technology capable of screening ULD pallets and containers, but according to officials, technologies to detect items deemed threats in pallets and containers are not expected to be available for several years.

[34] DHS deemed the reasons why passenger air carriers would have difficulty implementing additional security measures as sensitive security information. Thus, it was omitted from this report.

[35] See, e.g., 49 C.F.R. §§ 1544.105(b)(3); 1544.305(d); 1546.105(b).

[36] To support the requirement on industry to screen 100 percent of air cargo transported on passenger aircraft from U.S. airports, TSA developed the CCSP. The CCSP allows screening of domestic cargo at points throughout the supply chain by an approved Certified Cargo Screening Facility (CCSF) prior to the arrival of the cargo at the airport.

[37] GAO-10-446.

[38] Following the October 2010 incident, TSA implemented a new requirement to screen cargo transported on all-cargo aircraft for explosives using the same TSA-approved screening methods for cargo transported on passenger aircraft.

[39] Based on 2010 BTS data by freight configuration for all-cargo carriers. Passenger aircraft transported the remaining 33 percent (3.6 billion) of air cargo pounds.

[40] As of January 2012, the ACAS pilot included 4 of the 70 all-cargo carriers that service the United States and is focused on about 84 geographic locations.

[41] DHS, *DHS Policy for Integrated Risk Management* (May 27, 2010).

[42] GAO/AIMD-00-21.3.1.

INDEX